PROTOPLASMATOLOGIA
HANDBUCH
DER PROTOPLASMAFORSCHUNG

HERAUSGEGEBEN VON

L. V. HEILBRUNN UND F. WEBER
PHILADELPHIA   GRAZ

MITHERAUSGEBER

W. H. ARISZ-GRONINGEN · H. BAUER-WILHELMSHAVEN · J. BRACHET-
BRUXELLES · H. G. CALLAN-ST. ANDREWS · R. COLLANDER-HELSINKI ·
K. DAN-TOKYO · E. FAURÉ-FREMIET-PARIS · A. FREY-WYSSLING-ZÜRICH ·
L. GEITLER-WIEN · K. HÖFLER-WIEN · M. H. JACOBS-PHILADELPHIA ·
D. MAZIA-BERKELEY · A. MONROY-PALERMO · J. RUNNSTRÖM-STOCKHOLM ·
W. J. SCHMIDT-GIESSEN · S. STRUGGER-MÜNSTER

BAND II

CYTOPLASMA

D

VITALFÄRBUNG. VITALFLUOROCHROMIERUNG

2

THE METACHROMATIC REACTION

SPRINGER-VERLAG WIEN GMBH
1956

THE METACHROMATIC REACTION

BY

JOHN W. KELLY
RICHMOND, VIRGINIA

WITH 24 FIGURES

SPRINGER-VERLAG WIEN GMBH
1956

ISBN 978-3-211-80422-3   ISBN 978-3-7091-5529-5 (eBook)
DOI 10.1007/978-3-7091-5529-5

Softcover reprint of the hardcover 1st edition 1956

The Metachromatic Reaction[1]

By

JOHN W. KELLY

Department of Anatomy
Medical College of Virginia
Richmond, Virginia

With 5 Figures

Contents

[1] Certain statements in this paper refer to original work supported by grants-in-aid from the A. D. Williams Memorial Fund, Medical College of Virginia. and from the National Institutes of Health (G-4212), United States Public Health Service.

Introduction

Metachromasy exists when a pure dye stains a tissue section in a hue perceptibly different from the color characteristically associated with the dye. Thus, a dilute solution of toluidine blue is *blue*. Cell nuclei and certain basophilic components of the cytoplasm are stained in this color. A number of other histological elements are stained *red* by toluidine blue. The latter is the metachromatic color of the dye. Both extremes of color, as well as intermediate hues, may prevail in the same histological preparation.

The striking histological appearance of metachromasy is the most familiar and useful expression of the phenomenon. Colored drawings or photographs of various histological or cytological elements, stained by metachromatic dyes, have been published by GREEP (1954), JORPES (1946), KELLY (1950, 1954), MAXIMOW and BLOOM (1952), MICHELS (1938), PEARSE (1953), WISLOCKI, BUNTING and DEMPSEY (1947 b) and many others. It is a simple matter to demonstrate metachromasy. A dilute, aqueous solution of toluidine blue or Azure A will quickly stain cartilage, for example, an overall purple color. Upon dehydration and mounting, the cell nuclei retain a blue color which is easily distinguished from the metachromatic red or violet of the cartilage matrix.

Metachromasy is by no means restricted to histological preparations. If heparin is slowly added to a dilute solution of toluidine blue, the blue color of the dye solution becomes red, passing through a narrow violet range. Certain gels offer another *in vitro* demonstration of metachromasy. When an agar plate or agar grains are exposed to a dilute solution of toluidine blue, a deep purple or red reaction is seen in the gel. The sub-

strates, heparin and agar, are especially effective in eliciting the metachromatic reaction in a number of dyes.

To a degree, the metachromatic color is approached even in the *absence*
of a substrate, when an appropriate dye solution is concentrated. It can
be seen that a highly concentrated toluidine blue solution is distinctly
violet in contrast to the blue color of a dilute solution. The relation of the
"dilution shift" of many dyes to the histologist's metachromatic reaction
is an important one, to be discussed in a later section.

Metachromasy, in practice, has come to mean a special case of basic
dye interaction. Among the basophilic tissue constituents, there are some
that stain particularly intensely with non-metachromatic, basic dyes. These
same components often display metachromasy when exposed to a metachromatic, basic dye. The possibility of a comparable situation among
acidophilic tissue components and acid dyes has been suggested (BANK and
BUNGENBERG DE JONG 1939), though virtually nothing is known about acid
dye metachromasy. Unless otherwise specified, this discussion will refer
to the metachromasy of basic dyes, especially thiazine dyes. Even within
the thiazine group, usage has further selected toluidine blue and Azure
A as the best metachromatic dyes. The idea will be developed here that
many factors have favored the selection of a few useful dyes, leading
to the general association of metachromasy with basic dyes. Some of these
factors have nothing to do with the fundamental reaction and, indeed,
may have obscured certain aspects of the reaction.

A number of terms should be defined at this point. The *orthochromatic* color
of a dye is the "normal" color as it is seen in a dilute solution of the dye.
This is in contrast to the *metachromatic* color produced when the dye combines
with certain substrates. Any substance particularly effective in promoting the
metachromatic reaction of a susceptible dye is called a *chromotrope* (LISON 1935 a).
There is little choice between *metachromasia* and *metachromasy*, widely-used
names for the phenomenon. The former is more common among histologists and
the latter more common among those who make chemical studies of the reaction.
LEVINE and SCHUBERT (1952 a), preferring *metachromasy* themselves, note that
HOLMES (1926 a) first used this word in English and that *metachromasia* did not
appear until 1934, in translation from the French, and not until 1940 in the
original English (HEMPELMAN 1940). To these facts might be added the observation
that the original "Metachromasie" of EHRLICH (1877) has typically been translated
by many Europeans, LISON and MUTSAARS (1950) for example, into the English
metachromasy. Other terms, such as *metachromism* (NAGEL 1948) and *metachromatism* (CLOWES and OWEN 1904; RILEY 1953) enjoy little use today. There
are several terms used to describe staining phenomena which must be clearly
distinguished from metachromasy. *Polychromasia* is the appearance of two or
more colors in the same stained preparation by the use of special dye mixtures.
CONN (1953) discusses these mixtures, such as the Romanovsky-type stains, in some
detail. *Allochromasy* was used by MICHAELIS to describe the staining produced
by impure dyes or dye mixtures. This is not to be confused with the *allochrome*
procedure or allochroic color change of LILLIE (1952 b), defined with respect to a
special periodic acid-Schiff procedure. All of these—polychromasia, allochromasy, the
allochrome procedure—differ from metachromasy in their dependence on impure or
mixed dyes. The metachromatic reaction can be established in a pure dye solution.

The empirical use of metachromasy is well founded in microscopic work. Endowed with some histochemical meaning, it is one of the few straight dye reactions currently holding its position in histochemistry. From the great backlog of dye methods, there has been small yield in terms of objective, qualitative description or accurate quantitation. Yet modern histochemistry can ill afford to overlook the type of information which only dyes can provide (SINGER 1954). It is conceivable, for example, that dyes may lead us to knowledge of the "macromolecular structure of some important large molecule complexes *as they occur in the tissue and cells*" (POLLISTER and ORNSTEIN 1955).

Two specific examples will illustrate the manner in which valuable but empirical methods have been given new meaning. MICHAELIS, in 1900, discovered that Janus green B stained mitochondria selectively. To this day, the dye has been used as a specific reagent for mitochondria, both *in situ* and in homogenates. Quite recently, LAZAROW and his co-workers have found the explanation for this reaction. Janus green B is specifically re-oxidized by cytochrome oxidase, an enzyme found exclusively in mitochondria. References to the Janus green studies are found elsewhere in this Handbuch (LINDBERG and ERNSTER 1954). Another old method, developed around 1900 by UNNA and by PAPPENHEIM, is the methyl green-pyronin differential stain. BRACHET, in 1940, revived the method for the histochemical distinction of DNA and RNA. KURNICK (1952) and TAFT (1951) summarize the present status of the methyl green-pyronin stain as a histochemical procedure, including possible mechanisms and limitations. Like the Janus green and methyl green-pyronin methods, the metachromatic reaction offers both a useful technique and an exceptional opportunity to explore the mechanism of staining. There is no phenomenon in the histologist's arsenal more striking.

It is the intention here to display the metachromatic reaction in all its aspects. Considerable reliance is placed on a number of extensive papers containing references to the earlier literature (BANK and BUNGENBERG DE JONG 1939; BIGNARDI 1946; LISON 1935 a, 1936 a; MICHAELIS 1947; SYLVÉN 1954). General descriptions of metachromasy are found in books on histochemistry (GOMORI 1952; LISON 1953; PEARSE 1953).

The properties of metachromatic dyes and of chromotropes are separately examined, as well as the reaction between them. The histology of natural chromotropes involves their distribution among living forms, their localization down to the cellular level, and changes observed under normal and pathological conditions. To these chemical and histological studies, the practical usage of the reaction is an important adjunct. In turn, histological and especially chemical information has led to theories of metachromasy. Finally, the physiological importance of metachromasy, largely of the chromotropes themselves, is briefly discussed.

Historical Outline, 1875—1935

CORNIL (1875), HESCHL (1875) and JÜRGENS (1875) independently described a peculiar staining of amyloid. Certain triphenylmethane dyes, like methyl violet, dahlia (HOFMANN's violet) and crystal violet, stained amyloid in colors distinctly different from the ordinary colors of the dyes. It is curious that these observations, the first reported instances of metachromatic staining, should have been with triphenylmethane dyes and a chromotrope such as amyloid. The explanation for triphenylmethane metachromasy is even today less well-known than that for thiazine dyes and the chemical nature of amyloid, after all these years, is still unknown.

EHRLICH (1877), while still a medical student, found that mucin and certain granular cells in the connective tissue were stained by dahlia in the same way that amyloid was. The granular cells, probably first seen by VON RECKLINGHAUSEN and by KÜHNE as early as 1863 (MICHELS 1938), were called "Mastzellen" (EHRLICH 1879 b); they were stained "metachromatisch, d. h. in einer von dem angewandten Farbtone abweichenden Nuance" (EHRLICH 1879 a). Thus, EHRLICH defined and named both the metachromatic reaction and the mast cells. His is also the first list of metachromatic dyes published (EHRLICH 1877). A second list of metachromatic dyes was published by LISON (1935 a), along with a list of early discoveries of metachromasy. Other papers containing numerous references to early literature should also be consulted (BANK and BUNGENBERG DE JONG 1939; LEHNER 1924; MICHAELIS 1903, 1910, 1926; MICHELS 1938; v. MÖLLENDORFF 1924).

Amyloid, mucin and mast cells were thus the first chromotropes. Between 1875 and 1910, consistent with the rapid development and wide, sometimes sanguine, use of aniline dyes in all phases of histology and pathology, a number of other metachromatic tissue elements were discovered. These chromotropes fall roughly into three categories. The broadest classification must include metachromatic "granules" or "vacuoles" in the cells of bacteria, yeasts, fungi, algae and Protozoa (for references, see HENRICI, 1930, and GUILLIERMOND, MANGENOT and PLANTEFOL 1933). Metachromatic dyes were also useful in staining the matrix of bone and cartilage, though many of the methods used took advantage only of the intense basophilia of the matrix for such dyes (GATENBY and BEAMS 1950). An acid dye, indigocarmine, was found to be metachromatic with bone by KÖLLIKER in 1888 (see 1937 edition of GATENBY and BEAMS 1950). Finally, DUSTIN (1947) discusses a type of metachromatic vital staining that was seen in erythrocytes and reticulocytes. All of these instances of metachromasy will be considered in detail later.

A number of theories of metachromasy were advanced from about 1900 to 1930, all of them based on (a) histological observations or (b) experiments with solutions of dyes alone (CLOWES and OWEN 1904; HANSEN 1908; LEHNER 1924; MICHAELIS 1910, 1926; v. MÖLLENDORFF 1924; PAPPENHEIM 1906). LISON (1935 a) and MICHELS (1938) present excellent general summaries of metachromatic theories during this early period.

In 1935, an event was reported that offered an opportunity of unifying the histological findings on metachromasy with dye solution studies: LISON (1935 a, 1935 b, 1936 a), discovered that the metachromatic reaction could be reproduced in a test tube. A chromotrope, in minute quantities, brought about the familiar color change in a solution of dye. Based on his original

investigation of fifty-two dyes, a variety of chromotropes, and the influence of different agents upon the reaction, Lison's crucial statement was
that the "phénomène de métachromasie est en réalité lié à une constitution
chimique définie et constitue une véritable réaction histochimique spé-
cifique; elle est caractéristique des esters sulfuriques de substances à poids
moléculaire élevé." Regardless of certain qualifications subsequently
imposed upon Lison's statement and upon his criteria for metachromasy,
the importance of his discovery cannot be minimized. It is a fact that
all investigations of the metachromatic reaction since 1935, especially
where theory is involved, have directly or indirectly embraced some aspect
of the metachromatic reaction *in vitro*.

Dyes

Certain properties of metachromatic dyes can be discussed without
reference to a chromotrope. While the dyes most often used for metachromatic staining are not strikingly different from dyes in general,
knowledge of their structure and of their behavior in aqueous solution
facilitates the prediction of a metachromatic reaction before dye and
chromotrope are brought together. It is these differences in structure and
behavior that will be treated in this section, deliberately avoiding discussion on the general properties of dyes. A number of organic chemistry
books treat the relation of chemical structure and color adequately; that
of FIESER and FIESER (1950) is recommended. Much of the general information and the terminology used in this section came from BRODE (1949),
CONN (1953, GIBSON (1949), and MELLON (1948, 1950). These works are also
valuable for their extensive reference lists.

I. Optical Properties

A. Visual Observations

1. The dilution shift in aqueous solutions

HANSEN (1908) was the first to report that the more concentrated solutions
of certain dyes (e. g., thionine) exhibit hues different from those of dilute
solutions. This is a common observation for a number of colored compounds but it is especially striking with the metachromatic dyes. The
orthochromatic color of any of the dyes in Table 1 is ordinarily seen when
looking through a tube containing a dilute aqueous solution of the dye,
about 10^{-5} M or lower. With increasing concentration, *in the same tube,*
it is seen that the original hue lightens along with the expected increase
in density. LISON (1935 a) pointed out that the metachromatic shift is
invariably *hypochromic.* HOLMES (1926 a) stated that, while the concentration effect was displayed by many dyes not ordinarily considered metachromatic, all metachromatic dyes "assume their metachromatic colors when
their aqueous solutions are made sufficiently concentrated." HOLMES held

Table 1. *Metachromatic Dyes.*

These cationic dyes are reported to be metachromatic or to show significant color changes with tissue elements or chromotropes in solution. Anionic dyes are omitted from the list. The references cited will indicate variability in reaction or conflicting observations. Dye names are preferred by CONN (1953) unless doubt exists as to synonymy; in such cases the name is that given by the original author. Liberty has been taken in converting shades (e. g., crimson, pink, rose) to parent hues.

Abbreviations

B—blue, G—green, O—orange, P—purple, R—red, V—violet, Y—yellow, (F)— loses fluorescence, (S)—slight visual change or detectable spectrophotometrically.

Colour Index Number	Dye Group and Dye Name	Color Shift Orthochromatic	Color Shift Metachromatic	References
	Azine			
—	Heliotrope of tannin . .	RV	R	16
—	Indazine M	B	R	16
857	Magdala red	R	R, (?)	16
825	Neutral red	R, OR	O, PR, Y	1, 10, 16, 22
826	Neutral violet	V	R	1, 16
840	Phenosafranin	R	—	17, 19, 21
841	Safranin O	R	Y	16, 17, 22
	Azo			
331	Bismarck brown Y . . .	O	Y	10
135	Indole blue B	B	V	16
133	Janus green B	G	BR	1, 16
	Oxazine			
877	Brilliant cresyl blue . .	B, V	B, P, R	1, 9, 10, 11, 16, 22
—	Cresyl violet	B	P	10, 22
909	New blue B (R)	B	V	16
913	Nile blue sulfate	B, BG	BR, V	1, 10, 16, 17, 21
—	Oxonine	B	R	16, 20
	Phenylmethane			
681	Crystal violet	V	R	1, 2, 8, 11, 15, 16, 17, 21, 27
682	Ethyl violet	V	—	16, 22
677	Fuchsin, basic	R	RY, (S)	1, 10, 16, 17, 21
—	Glacier blue	BG	BV	16
679	Hofmann's violet . . .	V	R	4, 16
686	Iodine green	G	RV	28
657	Malachite green	G	—	1, 16, 17, 19, 21
680	Methyl violet (R, 2 R, 4 R)	V	R	11, 16
676	Pararosanilin	R	O, RY	11, 16
—	Sétocyanine O	BG	BV	16
689	Spirit blue	B	V	16
729	Victoria blue B	B	R	16

Colour Index Number	Dye Group and Dye Name	Color Shift		References
		Ortho-chromatic	Meta-chromatic	
	Phenylmethane			
728	Victoria blue R	B	R	16
—	Victoria blue 4 R . . .	B	P, R	16, 23
	Quinoline			
808	Pinacyanol	B	R	1, 3, 17, 18, 20, 24, 26
	Thiazine			
923	Azure A	B	R	6, 10, 16, 22, 29
923	Azure B	B	R	6, 16, 22
923	Azure C	B	R	6, 16, 22
922	Methylene blue	B	(S)	1, 6, 14, 15, 16, 19, 22, 25, 27
924	Methylene green	G	—	1, 14, 16, 27
927	New methylene blue N .	B, BV	R, V	7, 16
920	Thionine	B, BV, V	R	5, 6, 8, 10, 11, 16, 17, 22, 25
925	Toluidine blue O	B	R	1, 2, 10, 12, 14, 16, 17, 18, 19, 20, 22, 30
	Xanthene			
740	Acridine red 3 B	R	Y	16, 22
790	Acriflavine	Y	(F)	1, 10, 16, 19
739	Pyronin Y	R	(F)	1, 10, 16, 22

[1] BANK and BUNGENBERG DE JONG 1939; [2] CARNES and FORKER 1954; [3] CORRIN and HARKINS 1947; [4] EHRLICH 1877; [5] EPSTEIN, KARUSH and RABINOWITCH 1941; [6] HAYNES 1927; [7] HIGHMAN 1945; [8] HOLMES 1926 a; [9] HOLMES 1928; [10] JAQUES, BRUCE-MITFORD and RICKER 1947; [11] KELLEY and MILLER 1935 b; [12] KELLY 1955 b; [13] KOIZUMI and MATAGA 1953; [14] LEVINE and SCHUBERT 1952 a; [15] LEVINE and SCHUBERT 1952 b; [16] LISON 1935 a; [17] MERRILL and SPENCER 1948; [18] MERRILL, SPENCER and GETTY 1948; [19] MICHAELIS 1947; [20] MICHAELIS 1950; [21] MICHAELIS and GRANICK 1945; [22] MICHELS 1938; [23] OHUYE and MURAKAMI 1953; [24] PROESCHER 1933; [25] RABINOWITCH and EPSTEIN 1941; [26] SCHEIBE 1938; [27] SCHUBERT and LEVINE 1953; [28] STILLING 1886; [29] SYLVÉN and MALMGREN 1952; [30] WEISSMAN, CARNES, RUBIN and FISHER 1952.

little doubt that metachromatic staining and the solution phenomena had a common cause and demanded a common explanation.

Visual observations are seldom made on extremely concentrated solutions, since only the more dilute solutions will transmit enough light in ordinary vessels. However, it is possible to see the metachromatic color of a saturated solution of toluidine blue: a drop of solution between slide and coverslip forms a layer thin enough to permit observation of the metachromatic *red* color. Dried films of toluidine blue, examined by transmitted light, are reddish. While such experiments are easy to perform and crudely instructive, they must be carefully controlled in view of the characteristics of the human eye.

Organic solvents, acids, bases, salts, and variations in temperature modulate the fundamental metachromatic shift in one way or another. Visual observations associated with the operation of these agents will be described later when the underlying ab orption changes are examined.

2. Dichromatism and metachromasy

MELLON (1950) says that "solutions showing the phenomenon of two unequally changing absorption bands exhibit a different hue in thick layers than in thin layers, or in different concentrations at the same thickness." This phenomenon he calls *dichroism.* CLARK (1948) describes the effect of concentration on bromcresol purple in alkaline solution, stating that in "the more dilute solution the stimulus reaching the eye is such as to give the appearance of a *bluish-purple* but in the denser solution the stimulus gives the appearance of *red.* For CLARK, this is the *dichromatic* effect.

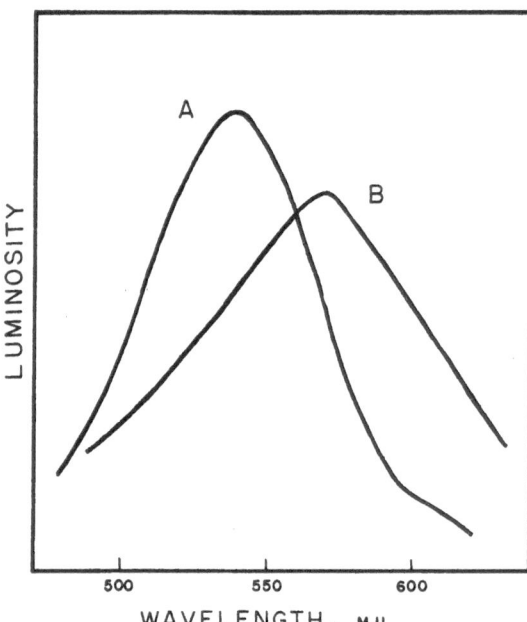

Fig. 1. Luminosity curves of toluidine blue. These curves are calculated from the relation, $E_c Tl$, where E_c is *relative energy* of the light source (I. C. I. illuminant C, approximating daylight), T is *transmittancy* of the solution and l is *relative luminosity* (I. C. I. standard observer). Luminosity is therefore an expression of the effective retinal stimulus. Curve A represents toluidine blue alone; curve B represents a solution of toluidine blue plus heparin. (Concentrations of dye and chromotrope are the same as in Fig. 3.)

Both of these terms have been used in other connections. Dichroism is a definite property associated with oriented particles when observed by polarized light (BEAVEN, HOLIDAY and JOHNSON 1955). It is not observed in ordinary solutions except under conditions of flow. Dichroism, by this definition, is not restricted to the visible region. In the visible region, however, "the combination of a dye with an oriented macromolecule may lead to orientation of the dye molecules, causing the visible light absorption of the dyed macromolecule to show dichroism" (BEAVEN, HOLIDAY and JOHNSON 1955). Dichromatism describes the state of being able to see only two of the three primary colors; it is a form of color-blindness (ADLER 1953). Dichroism is well-established as a physical description of oriented systems examined by polarized light. *Dichromatism* seems the better term to use for that phenomenon involving concentration of a solution, depth of absorbing layer, and characteristics of the human eye. This is quite distinct from the use of the word to characterize a color-vision defect.

Dichromatism and metachromasy are visual phenomena that cannot be distinguished without recourse to objective description. The most important concept for this distinction is the *photopic* visual characteristic of the human eye (color- or cone-vision). The color vision of the "average observer" may be described in terms of a *relative luminosity* curve (see CLARK 1948. p. 646; MELLON 1950, p. 119), showing the maximal sensitivity of the eye to radiant energy at 555 mμ, i. e., to yellow-green light. Using the relative luminosity curve, it is then possible to derive *luminosity* curves which express the ultimate retinal stimulus of systems differing widely in radiant power available at the source and in transmittance of the absorbing solution (Fig. 1).

Dichromatism is a broad term referring to any change of hue with varying concentration or thickness of a solution. Both non-metachromatic and metachromatic solutions may exhibit dichromatism; only photometric measure-

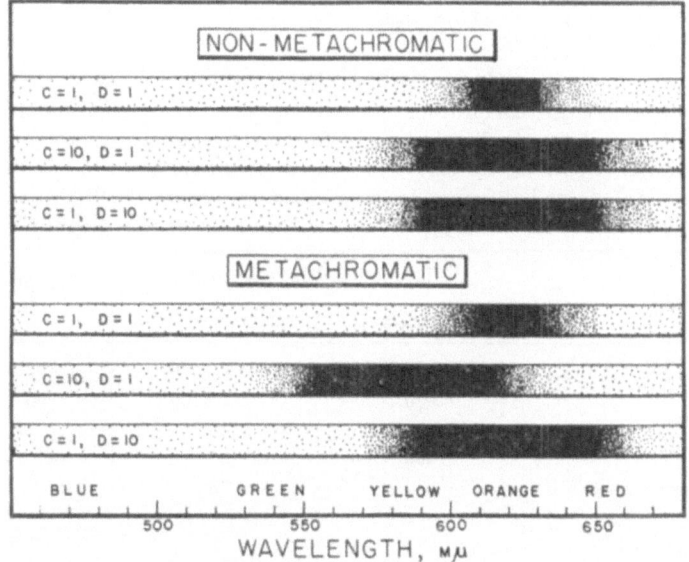

Fig. 2. Distinction between *metachromasy* and *dichromatism*, illustrated by arbitrary spectra of blue dyes. Regions of absorbance maxima are shown as black areas. Concentrations *(C)* and absorbing-layer depths *(D)* are relative only. The bands are broadened by increasing either dye concentration or depth of layer, leading to slight changes in hue. The visual effect of these slight changes increases as they encroach on the region near 555 mμ. It is, however, the marked shift of the absorbance band toward lower wavelengths with increasing concentration that distinguishes the *metachromatic* from the *non-metachromatic* dye. Both dyes partake of *dichromatic* effects.

ment will show that the spectral basis for dichromatism differs somewhat for the two types of solutions. The following points are emphasized:

(1) A change in hue necessarily involves the photopic vision of the human eye, whether the visual effect is related to a system changing in depth or in concentration.

(2) A change in thickness of absorbing layer always leads to broadened absorption bands, never to a shift of absoption bands (in ordinary solutions). This effect is common to metachromatic and non-metachromatic solutions.

(3) An increase in concentration also leads to broadened absorption bands; this may or may not be accompanied by a wavelength shift. If a wavelength shift occurs, the system is metachromatic.

Fig. 2 is an attempt to present the elements of this distinction in a graphic way.

3. Other visual factors

Certain additional factors must be considered when visually observing solutions and stained sections, especially the latter. Not all of these factors are independent of dichromatism. They may involve technical aspects of microscopes and lighting systems. Considering the differences between early and modern microscopy, some of these factors are perhaps of historical interest only. GAGE (1943) should be consulted for details.

Chromatic aberration of microscope optics is considerably less today than it was seventy-five years ago. Anyone who has examined objects

Table 2. *Metachromatic Shifts of Common Dyes.*

The value of a dye for metachromatic staining purposes is determined by the *magnitude* of the shift of absorbance to lower wavelengths in relation to the range of *maximal sensitivity for color stimulation* of the human eye (about 555 mμ). Failing to satisfy one or both requirements, those dyes marked with an asterisk (*) are relatively poor metachromatic stains. (For references, see Table 1.)

Dye	Absorbance Maxima (mμ)		Magnitude of Shift (mμ)
	Orthochromatic	Metachromatic	
Toluidine blue	630	480—540	90—150
Azure A	620	480—530	90—140
Pinacyanol	600	480	120
*Methylene blue	665	570	95
Brilliant cresyl blue . .	625	530	95
New blue R	622	533	89
Crystal violet . . .	590	510	80
*Basic fuchsin	543	495	48
*Thionine	597	557	40

through an antique microscope will have noticed annoying "color fringes," particularly associated with very small objects. Furthermore, objects whose diameter approaches the theoretical resolution of the light microscope (about 0.2 μ) tend to be *reddish* in hue.

Early microscopists used first sunlight and later the flame as light sources, before the advent of electric lights. Sunlight is, of course, highly variable and the flame offered some measure of constancy. The light of an ordinary flame, however, is dominantly *red*. Even the tungsten lamp is rich in energy of longer wavelengths.

While we cannot question many of the earlier reports of metachromasy, especially since improved microscopic and photometric equipment have enabled their confirmation, there is some uncertainty associated with many such reports. This is particularly true for the small "granules" and "vacuoles" abundantly found in bacteria, yeasts, fungi, Protozoa and other cells.

It is probably no accident that the "best" metachromatic dyes are blue, shifting toward reddish metachromatic colors. Given dyes of various colors, with metachromatic shifts of equal magnitude, the eye will best distinguish those color changes based on absorption changes in the region of yellow-green light, i. e., close to 555 mμ. MELLON (1950) says that the intensity of deep blues and deep reds cannot be duplicated within fifteen to twenty per cent, while greens and oranges can be duplicated within two per cent. Since it is *transmitted* light to which the eye is sensitive. dyes whose metachromatic shifts involve transmittance changes across or into the yellow-green will provoke the maximal retinal stimulus. Table 2 shows corresponding absorbance changes for several metachromatic dyes.

B. Photometric Observations

1. Absorption laws

a) Beer's law

Beer's law is conveniently stated in the form,

$$A = k' c$$

which shows the direct relation between absorbance, A, and the concentration, c, of absorbing material, for a constant cell depth. The constant. k', varies with wavelength. The familiar plot of absorbance against concentration is a straight line for compounds which do not deviate from Beer's law. The specific absorbance of some metachromatic dyes does not reach a constant value above 10^{-6} or 10^{-7} M (MELLON 1950; RABINOWITCH and EPSTEIN 1941).

Deviations from Beer's law were recorded as early as 1908 (SHEPPARD and GEDDES 1944 a). It was later found that Beer's law deviations were extremely common among organic dyestuffs (HOLMES 1924). HOLMES (1926 a) was the first to relate Beer's law deviations to metachromasy. Deviations from Beer's law, it is generally agreed, are probably more common among organic dyes in aqueous solution than is strict adherence to the law (MICHAELIS and GRANICK 1945). SHEPPARD and GEDDES (1944 a) consider such deviations to be part of a general pattern, "of which one extreme variant —of zero order—is conformity to Beer's law."

b) Bouguer-Lambert law

There are no known deviations from the Bouguer-Lambert law,

$$A = k'' d$$

for compounds in solution, where orientation is not ordinarily found. The law says that, for a given concentration of absorbing constituent, absorbance is proportional to the depth of the absorbing layer, k'' being a constant which varies with wavelength. A strictly linear relation is obtained by plotting absorbance of a solution against cell depth, at a given concentration. The Bouguer-Lambert and Beer's laws are often combined in a useful form of the general absorption law:

$$A = k c d$$

It is notable that, if deviations from the Bouguer-Lambert law were common, dichromatism would assume a more complex character. Distinctions between metachromasy and such dichromatism would be correspondingly difficult. The importance of Bouguer and Lambert's law is discussed again when the microspectrophotometry of tissues is considered.

2. The dilution shift in aqueous solution

The changes in hue when a metachromatic dye solution is concentrated are clearly related to known spectral changes. Many authors present absorption spectra illustrating the dilution shift (HOLMES 1924; MICHAELIS 1947; RABINOWITCH and EPSTEIN 1941; WEISSMAN, CARNES, RUBIN and FISHER 1952). There are several features common to all these curves. In dilute solution, a metachromatic dye may exhibit only one major absorption band, the α-band, with perhaps a shoulder toward shorter wavelengths. With increasing concentration, the shoulder is developed at the expense of the α-band, leading to a distinct β-band when the concentration is sufficiently high. β-bands lie about thirty to sixty millimicrons lower than the corresponding γ-bands. Ordinarily, the highest concentrations of dyes studied is of the order of 10^{-3} or 10^{-2} M; even these concentrations require absorption cells 0.01 cm. deep (RABINOWITCH and EPSTEIN) and it becomes virtually impossible to accurately calibrate cells thinner than 0.005 cm. deep (MICHAELIS and GRANICK 1945; MICHAELIS 1947). The occurrence of the γ-band, which arises with the addition of a chromotrope at even lower wavelengths than the β-band, is therefore not usually observed in solutions of dye alone. Several authors have studied extremely concentrated solutions (BRODE 1955; CORRIN and HARKINS 1947; SCHEIBE 1938) or dyes in the solid state (HOLMES 1926 a; KELLEY and MILLER 1935 b) and they find a γ-band.

The terminology used here and by other authors is that of MICHAELIS (1947). There is some confusion following MICHAELIS' (1950) last paper, when he re-named the γ-band, calling it the "metachromatic or μ-band." This was apparently in deference to the spectral complexity of a dye like pinacyanol. CORRIN and HARKINS (1947) investigated the metachromasy of pinacyanol with soaps, clearly defining four bands: α- and β-bands at 615 mμ and 570 mμ respectively, an unnamed band at 550 mμ appearing in relatively concentrated solutions, and the γ-band at 480 mμ observed only in dilute solutions of soaps. MERRILL, SPENCER and GETTY (1948) found pinacyanol, in the presence of various silicates, to have absorption maxima at 630 mμ (α_1), 587-599 mμ (α), 546-550 mμ (β), and at 488-502 mμ (γ). MICHAELIS (1950) says pinacyanol displays the α- and β-bands in dilute solutions and a "γ-band" at higher concentrations. In solutions so concentrated as to have the properties of a liquid crystal (SCHEIBE 1938) and also in agar, MICHAELIS notes a new band at 640 mμ (unnamed). Still a fifth band, the μ-band, is said by MICHAELIS to arise at a very much shorter wavelength than the γ-band previously described by other authors. This is the point of confusion, since the lowest band shown in MICHAELIS' figures is precisely at 480 mμ, which is the γ-band of CORRIN and HARKINS. The case of pinacyanol is somewhat different from other dyes, where there is no good reason for abandoning the original use of "γ-band."

The manner in which the metachromatic shift progresses is important. In curves illustrating the conversion of α-band to β-band with increasing dye concentration, RABINOWITCH and EPSTEIN (1941) show no *isosbestic* point. The isosbestic point is characteristic of equilibria between two species, for example, pH-indicators (FORTUNE and MELLON 1938; BRODE 1955) or cis-trans equilibria (Brode 1955). Rabinowitch and Epstein's investigations were in relation to assumed equilibria between monomeric and dimeric forms of methylene blue and thionine, accounting respectively for the α-bands and β-bands (called "M-" and "D-bands" by these authors). Holmes' (1924) curves show a decline in the main peak and rise in a new peak at some lower wavelength, with increasing dye concentration. Lison (1935 a) thought this was because Holmes studied only extreme concentrations, his own curves showing a simultaneous lowering of the main band and displacement toward lower wavelengths. Since Lison's curves did not show the isosbestic point, he felt this spoke against the idea of tautomerization. Kelly (1955 b) did not observe the isosbestic point with toluidine blue in the presence of heparin or chondroitin sulfate, while Jaques, Bruce-Mitford and Ricker (1947), using Azure A, found an isosbestic point at 555 mμ for a family of absorption curves of the dye with different concentrations of heparin.

3. Organic solvents

The spectra of metachromatic dyes in alcohol, acetone and other organic solvents resemble those in very dilute aqueous solution (Lison 1935 a). The α-band is somewhat higher in alcohol—it may be displaced according to the nature of the solvent—and there is never a β-band: Beer's law is obeyed (Michaelis 1947). This may be due to an insulating effect of organic solvents, where the dye molecule is almost completely solvated ("homologated" to the solvent), while in water only the relatively sparse and isolated hydrophilic groups will be solvated, increasing the probability of organic aggregates (Sheppard and Geddes 1944 a). Ion aggregates in organic solvents have little effect on absorption spectra, say Sheppard and Geddes, while molecular aggregates occurring in water led to striking effects on spectra.

4. Influence of other agents

a) pH

Acids suppress metachromasy (Lison 1935 a). To the eye, this appears as a deepening of hue; the *blue* of a dilute solution of toluidine blue appears *blue-green* (Bank and Bungenberg de Jong 1939). Epstein, Karush and Rabinowitch (1941) show the absorption spectra of thionine (2.5×10^{-5} M) over a wide range of concentrations of hydrochloric acid. In 0.1 M acid, the dye has a spectrum similar to that in water alone, with a peak at 600 mμ and a shoulder at 560–570 mμ. In 4.8 M acid, the new peak is at 670 mμ with a shoulder at 600 mμ. It is interesting that an isosbestic point occurs at about 614 mμ. Kelley and Miller (1935 b) had previously determined the peaks of thionine, at pH 1, to be 670 mμ, 608 mμ, and a trace at 560 mμ. They thought the new peak at 670 mμ indicated the existence of a new dye, the dihydrochloride. The isosbestic point of Epstein, Karush and Rabinowitch indicates an equilibrium between the hydrochloride and dihydrochloride of thionine.

Certain cationic dyes (e. g., methyl green) are doubly-charged at neutral

or only slightly acid reaction, where most thiazine, oxazine and triphenylmethane dyes are singly-charged (MICHAELIS 1950). Beer's law is valid for
doubly-charged cations, less valid for singly-charged cations. This is a
possible explanation for adherence of a dye like methyl green to Beer's
law, which is not obtained for many other cationic dyes except at extremely low pH.

The addition of base gives rise to colors that may closely resemble the
metachromatic colors of a number of dyes (LISON 1935 a; BANK and
BUNGENBERG DE JONG 1939). It will later be shown that this is due to the
formation of one of the dye bases, whose absorption spectra can usually
be distinguished from those characteristic of metachromasy.

b) Salts

Salts generally influence the color of metachromatic dye solutions like
acids and organic solvents; they suppress metachromasy and are never
true chromotropes (LISON 1935 a). BANK and BUNGENBERG DE JONG (1939)
defined metachromasy as *any* appearence of new bands with the concurrent
loss or intensity change of the bands characteristic of the dye in aqueous
solution. On this basis, they present a table of twenty-five salts, showing
the effect of inorganic anions as "metachromators" with toluidine blue.
The effect of salts is further discussed by MICHAELIS (1947). A dye may be
salted out or remain in a relatively stable colloidal state, depending on the
concentration and nature of the salt. The non-metachromatic dyes tend to
precipitate as crystals with the addition of salts, while the metachromatic
dyes resist precipitation up to high salt concentrations and may remain in
colloidal solution. The absorption characteristics of such solutions approach
those of metachromatic solutions. The spectra of dyes showing no β- or
γ-bands, like phenosafranine, are only depressed by the addition of neutral
salts, not displaced.

c) Temperature

Temperature has a pronounced effect on pure dye solutions, increased
temperatures generally suppressing metachromasy (LISON 1935 a; BANK and
BUNGENBERG DE JONG 1939). Heating a tube of toluidine blue in dilute
aqueous solutions produces a barely perceptible shift toward *greenish* hues:
cooling in an ice-bath has the opposite effect. Spectrally, increased
temperature favors the establishment of the α-peak at the expense of the
β-peak. This thermochromic effect is illustrated for thionine at high concentrations (3.5×10^{-3} to 3.5×10^{-2}) over the range 10–70° C. (BRODE 1935).
Dilution and increased temperature have the same effect, which may in
some cases represent an equilibrium between associated and non-associated
forms of the dye, in other cases a change in molecular configuration. BRODE
shows no isosbestic point between the two peaks involved (at about 600 mμ
and 550 mμ); the curves never cross.

The effects of temperature, as well as those of organic solvents, acids
and salts, are completely reversible. The effect of bases is not reversible
(BANK and BUNGENBERG DE JONG 1939).

5. Ultraviolet absorption spectra

It has been stated that the ultraviolet absorption bands of many dyes are "much less subject to variation of conditions" than those in the visible region (MICHAELIS 1947). They are said to be practically independent of concentration even when visible bands deviate from Beer's law and they are not affected by the presence of a chromotrope (agar) (MICHAELIS and GRANICK 1945). In short, "the ultraviolet bands are, for all practical purposes, unaffected by all those circumstances which so strongly influence

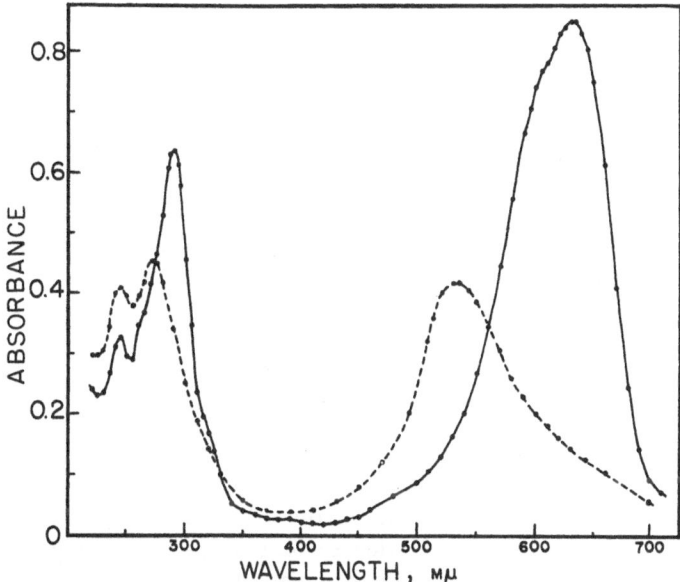

Fig. 3. The metachromatic shift of toluidine blue. Toluidine blue has absorbance maxima at 630 mμ (α-peak), 290 mμ, and 245 mμ *(solid line)*. Dye concentration, 2.45×10^{-5} M. Addition of a chromotrope—heparin, 4.0×10^{-5} eq./1—induces a new metachromatic peak at about 535 mμ (γ-peak) and the color is now red *(dotted line)*. Note also that the 290 mμ peak shifts to 275 mμ, while the 245 mμ peak is only lowered. (Beckman spectrophotometer, model DU, 1.0 cm. silica cells.)

the absorption in the visible spectrum" (MICHAELIS 1950). Yet SCHUBERT and LEVINE (1953) found very interesting alterations in ultraviolet absorption spectra of methylene blue in the presence of chondroitin sulfate and hyaluronate. Methylene blue has five absorption bands in the visible and ultraviolet regions, at 665, 610, 570, 290, and 245 millimicrons. All of these were modified to some extent by chondroitin sulfate; with hyaluronate, all bands were affected except the 245 mμ band. Crystal violet, with bands at 590, 550, 510, 360, 305 and 250 millimicrons showed no change at all in the ultraviolet region in the presence of the same chromotropes, though all three visible bands were considerably affected. The same effect was later reported by KELLY (1955 a) for toluidine blue and heparin, where the 290 mμ peak of the dye displayed a "metachromatic" shift and the 245 mμ peak was unaffected (Fig. 3).

Even though the meaning of the ultraviolet shifts is obscure, they offer a new area for exploration. MICHAELIS (1950), even though denying the existence of Beer's law deviations in the ultraviolet, recognized the importance of ultraviolet bands for theoretical reasons. "The finding of variations in ultraviolet band intensities opens a new aspect of metachromasy that may help in the building up of the underlying theory" (Schubert and Levine 1953).

II. Chemical Properties

The classification, terminology and formulas of dyestuffs used here and in Table 1 are essentially those of Conn (1953). In Table 1, the name of a dye is usually the preferred name as given in Conn, unless some doubt exists as to the identity of the actual dye cited by an author and its synonyms. In this case, the cited name is given.

An attempt is made to avoid, as much as possible, discussion of the general chemical properties of dyestuffs, common to metachromatic and non-metachromatic dyes. It is more fitting to consider those properties which seem related to the behavior of the dyes which have come to be called metachromatic. Dye structure goes only part of the way in explanation of such behavior but it forms an important preliminary to consideration of the dye substrate—the chromotrope—and the metachromatic reaction.

A. Metachromatic Dyes

1. Cationic dyes

In practice, the cationic (basic) dyes are the important dyes from the standpoint of metachromasy. The chromogen of these dyes is an organic cation, whose auxochrome is invariably the amino-group ($-NH_2$) or a substituted amino-group. Thus, like other cationic dyes, the stainable substrate for such metachromatic dyes ordinarily must contain anionic groups.

a) Chromogen nucleus

Most classifications of dyes are based upon the chromophores (e. g., $-N=$, $-N=N-$) responsible for color. Since chromophores are arranged

Azine   Oxazine   Thiazine   Xanthene

Azo   Quinoline (a pinacyanol)   Triphenylmethane

in certain common relations to aromatic rings, it is most useful to classify dyes according to the resulting organic nucleus or *chromogen.*

Phenylmethane and thiazine dyes were the first known to exhibit metachromasy (p. 6). Certain dyes of these two groups are still the most important metachromatic dyes. Azine, oxazine, and xanthene dyes were subsequently added to the list (BANK and BUNGENBERG DE JONG 1939; LISON 1935 a; MICHAELIS 1903, 1910). Fewer, azo (JAQUES, BRUCE-MITFORD, and RICKER 1947) and quinoline (PROESCHER 1933) dyes are known to be metachromatic. The configurations of these chromogens are shown above. (p. 18).

b) Substituents

The substituent group invariably found in cationic metachromatic dyes is the amino-group ($-NH_2$) which constitutes the characteristic auxochrome of these dyes, responsible for their basic character.

Like all dyes, the color of a metachromatic dye is generally deepened by the introduction of alkyl groups or aromatic rings into (a) the nucleus or into (b) an amino-group. HAYNES (1927 a) examined this bathochromic influence in the case of thiazine dyes. She found that primary (α) and secondary (β) absorption bands were shifted toward longer wavelengths with alkylation, correlated with generally increased intensity of staining. However, the metachromatic staining quality of a dye decreased with alkylation in favor of orthochromatic staining. KELLEY and MILLER (1935 b) thought that alkylation increased the magnitude of the metachromatic shift, though attachment of phenyl groups to amino-N or to central nitrogen atoms greatly represses the shift in azine, oxazine and thiazine dyes.

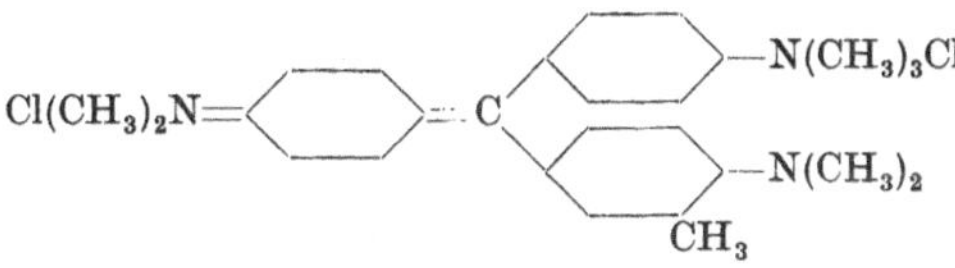

Iodine green, C. I. 686

STILLING (1886) felt that his results with the popular amyloid stain, methyl violet, were unsatisfactory and unreliable. Noting that another investigator (CURSCHMANN) had used methyl green for this purpose, STILLING adopted the use of iodine green. Presumably, the assumptions were that iodine green was methyl green whose methyl chloride has been replaced by methyl iodide and that an iodide might be a better amyloid stain because *iodine* itself colors amyloid in some unexplained manner. The formula of iodine green is shown above. If STILLING's dye contained iodide, this does not necessarily mean he was using what is today called iodine green, a trivial name having nothing to do with iodine. It may be that STILLING's dye was an impure dye, like crystal violet (CONN 1953) or methyl green (KURNICK 1950); iodine green would be even more susceptible to air oxidation than either of these related dyes.

2*

c) Non-chromogen radical

Cationic dyes are usually chlorides (hydrochlorides), less frequently acetates, iodides, sulfates or thiocyanates. The anion apparently has little effect on the metachromatic properties of the dye. Kelley and Miller (1935 b) found no difference among hydrochloride, acetate or iodide salts of a number of metachromatic dyes. Toluidine blue and other thiazine dyes are usually marketed as zinc chloride double-salts. Di Berardino (1954) observed a certain metachromatic reaction associated with purified toluidine blue which was not obtained when a certified, zinc-containing sample was used. She was able to rule out zinc chloride as an inhibiting factor.

2. Anionic dyes

Little or nothing is known concerning the metachromasy of anionic (acidic) dyes, whose color resides in the negatively charged chromogen. Chromophores in these dyes are usually the same as in the cationic dyes, the same classification being used for both groups. It is often only the auxochrome which serves to distinguish an anionic dye from a cationic dye. For most dyes, the auxochrome is the sulfonic ($-OSO_2^-$) group.

In the few studies on "metachromasy" of anionic dyes, the chromotropes have quite naturally been basic substances. The visual or photometric changes observed have been slight compared to those of cationic dyes.

Eosin Y, C. I. 768   Indigocarmine, C. I. 1180

Trypan blue, C. I. 477

Indigocarmine I a was found to be metachromatic with quinine and other bases (Bank and Bungenberg de Jong 1939). Other anionic dyes displayed spectral changes which are not metachromatic in the ordinary sense: Evans blue, Niagara sky blue, Niagara sky blue 6 B, and Trypan blue, all diazo dyes, showed shifts of about twenty millimicrons toward the red end of the spectrum, in the presence of plasma proteins (Rawson 1943). This indicates a "dampening of bond-energies of the dye as binding with protein takes place," an effect opposite that presumed to occur in the binding of cationic dyes to anionic chromotropes. Eosin, fluorescein, sky blue FF, and *acidified* 2, 6-dichlorindophenol displayed significant shifts with cationic detergents, some dyes showing both color changes and loss of fluorescence (Corrin and Harkins 1947). Whatever the relation of such color changes to our present idea of metachromasy may prove to be, "anionic analogs to the cationic metachromatic dyes have received scarcely any attention" (Levine and Schubert 1952 a) and must be exploited for possible contributions to the theory of metachromasy.

Since little is known about anionic metachromatic dyes, it cannot be said with certainty that the non-chromophore radical is without influence. The anionic dyes are typically sodium salts, though they may be ammonium or potassium salts as well. For dyes in general, the effect of the inorganic cation is unimportant.

The formulas of a number of anionic dyes that exhibit a type of "metachromasy" are shown above. (p. 20).

B. Non-Metachromatic Dyes

It is important to know which dyes, among those that have been investigated, are *not* metachromatic with any chromotrope. Even among common metachromatic dyes, it is known that a given dye may be metachromatic with several distinct classes of chromotropes. Conversely, a single dye group may show selectivity for chromotropes, exhibiting metachromasy with one group and not with another.

A large number of dyes are clearly not metachromatic, insofar as wide study has not revealed any evidence of their metachromasy. Among such dyes are the natural dyes (indigo, cochineal, orcein, hematoxylin and others), the diazonium-, tetrazonium- and tetrazolium salts, fluorans, phthaleins, nitroso and nitro dyes, and the mineral pigments. Indigocarmine is an exception among natural dyes (p. 20). The so-called "hematoxylin metachromasy" is a pH-effect (Kelley and Miller 1935 a).

Within a closely-related group of dyes, some may not be metachromatic. Lison (1935 a) made a detailed study of this question, drawing the following pertinent conclusions:

(1) Only those dyes are metachromatic which possess at least one unsubstituted amino-group.

(2) A metachromatic dye must be able to yield the imine base ($=NH$) and this base must have a color different from the orthochromatic color of the dye salt.

Methylene blue and Capri blue, with amino-groups fully substituted, are not metachromatic (Michaelis 1926). Lison makes the comparison among thionine, methylene azure, and methylene blue:

Thionine (metachromatic)   (*imine* base)

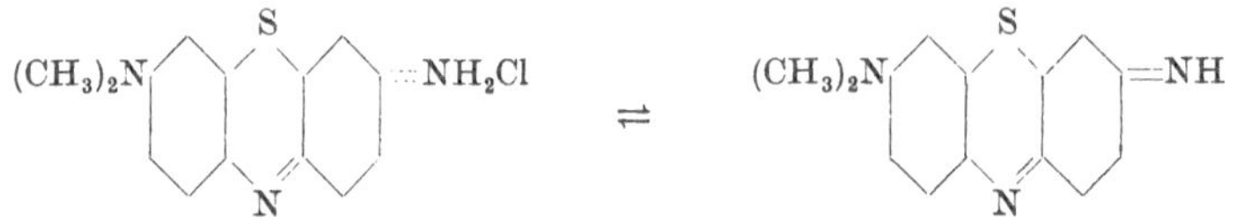

Methylene azure (metachromatic)   (*imine* base)

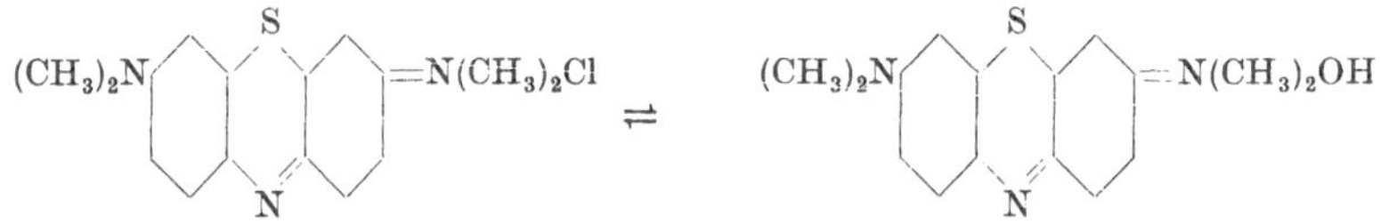

Methylene blue (non-metachromatic[2])   (*ammonium* base)

While LISON was able to extend the imine-base hypothesis to a large number of dyes (Table 1), some apparent exceptions were found. The azo dyes, Janus green and Janus blue, are metachromatic even though their amino-groups are fully substituted. In such dyes, says LISON, the amino-groups are substituted by the azo group and thus "return to the category of the azines." Furthermore, not all dyes capable of yielding

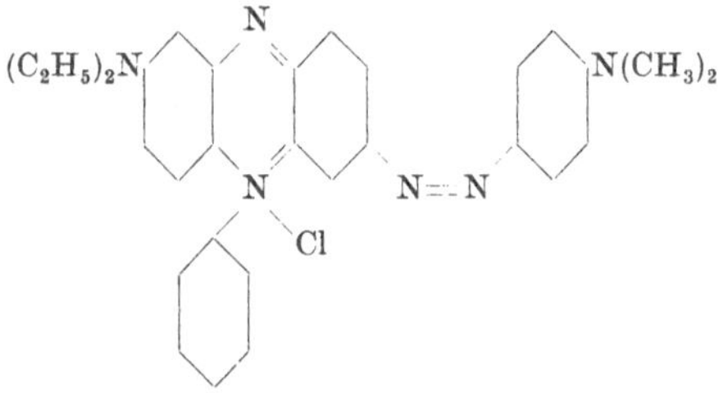

Janus green B, C. I. 133

an imine base are metachromatic (LISON 1935 a). Rhodamine B, according to LISON, possesses the appropriate chemical structure but is not meta-

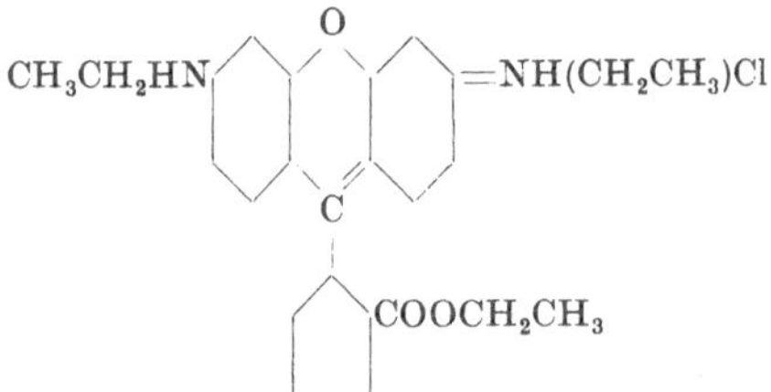

Rhodamine 6 G [3], C. I. 752

chromatic. It is said that the colors of this dye salt and its imine base are practically indistinguishable.

---

[2] See p. 47.

[3] According to CONN's (1953) formulas, rhodamine 6 G (corrected above) would yield the imine base, rhodamine B would not. LISON found none of six rhodamines to be metachromatic.

Continuing the discussion of dye bases, LISON disputes the use of "free base" and prefers the more specific "imine base," on the ground that certain dyes or dye groups are capable of yielding more than one base. He cites basic fuchsin, a triphenylmethane dye, as an example.

Basic fuchsin, *red*
(formula acc. to LISON)

C. Impure Dyes

The role of impurities in the history of metachromatic staining must be a large one. Some idea of the vague manner in which beautiful colors were observed with mysterious mixtures may be obtained by studying a book like that of GATENBY and BEAMS (1950; see especially earlier editions). Even today, many common biological stains are mixtures (CONN 1953). The work of the Biological Stain Commission (U. S.) has greatly improved the situation, though it must never be assumed that a certified stain is a pure stain.

1. Special dye mixtures

There are a number of dye mixtures which are deliberately prepared in order to obtain a desired staining effect. The most important of these mixtures, in relation to metachromasy, are the Romanowsky-type (GIEMSA, LEISHMAN, WRIGHT) so valuable in blood work. CONN (1953) presents an excellent discussion of such mixtures. Briefly, the basis for all these mixtures is the intentional oxidation of methylene blue to yield *polychrome methylene blue*, an indefinite mixture of azures. In various modifications of the original Romanowsky stain, the attempt is made to prepare the desired mixture by starting out with one or more of the supposed end-products or by controlling the oxidation more carefully. In any case, the staining produced by such mixtures includes colors that are definitely metachromatic. It is obvious that interpretations of such metachromatic reactions cannot be made.

2. Oxidized dye solutions

The "aged" or "ripened" solutions called for in many staining procedures often represent mixtures of the original dye and its oxidized homologs. These homologs tend to have orthochromatic colors which resemble the metachromatic colors of the original dye (cf. methylene blue and the azures) so that any "metachromatic" staining obtained with such mixtures might actually be a type of differential staining.

Hoffmann's violet and methyl violet, so frequently used by early microscopists, are known today to be mixtures (Conn 1953). Crystal violet, recommended for the staining of amyloid, is considered to be a definite compound; yet two of its synonyms are methyl violet and gentian violet which are mixtures of pararosanilin derivatives. Methyl green is not a metachromatic dye. It is known to contain variable amounts of methyl violet or crystal violet (Kurnick 1950; Taft 1951), yet Conn (1953) says that "pure methyl green may not always be desired by the biologist, as the dye owes part of the metachromatic properties for which it is prized to the presence of small amounts of the violet compound"!

There are few accurate statements on the storage of dye solutions. A purified sample[4] of toluidine blue was made up in glass-distilled water, 10 mg. of dye per liter, stored in a Pyrex bottle and kept in the ordinary light of the laboratory with no special precautions other than cleanliness. This stock was used to prepare a large number of dilute solutions over a period of seven months. Absorption readings on these dilute solutions at 540 mμ and 620 mμ declined only about three per cent in this time; occasional full spectra showed no change. Deterioration of thionine in Pyrex bottles is significantly less than in soft glass bottles (Epstein, Karush and Rabinowitch 1941).

3. Removal of impurities

The use of a dye not certified by the Stain Commission is to be discouraged, unless analysis of the dye is available. The dye-content stated for certified stains is reasonably accurate. Weissman, Carnes, Rubin and Fisher (1952) use Kjeldahl-N as a measure of dye-content for toluidine blue.

Recrystallization is often used to purify a dye. Levine and Schubert (1952 a) use a simple recrystallization method for thiazine dyes. It is not known if such techniques produce a dye free of closely related homologs or of non-dye impurities (e. g., zinc chloride).

Extraction of aqueous dye solutions with organic solvents has been widely used. By this means, some of the azures have been removed from methylene blue (Holmes 1927 a; Lasfargues and di Fine 1950) and crystal violet or closely related dyes have been extracted from methyl green (Kurnick 1952; Taft 1951).

Paper or column chromatography has been employed for the fractionation of several metachromatic dyes. Azure A, thionine and toluidine blue were investigated by this means, followed by spectrophotometric and histological studies of the whole dyes and their fractions (Ball and Jackson 1953; Kramer and Windrum 1955). None of the samples were

---

[4] Abbott Laboratories, Lot 8441—A sample of this dye was obtained from Dr. Henry G. Kupfer, Director of Clinical Laboratories, Medical College of Virginia. The dye was a gift to Dr. Kupfer from the Abbott Laboratories. Its dye-content was 92%.

homogeneous. The various fractions either had no staining ability, stained orthochromatically, or stained metachromatically. The most important finding of these authors is that a single, presumably homogeneous fraction may be recovered which stains metachromatically, even though whole dyes are mixtures in every case. Victoria blue 4 R has been purified by paper chromatography (OHUYE and MURAKAMI 1953).

Chromotropes

Reports on the occurrence of metachromasy and the detailed observations that microscopists have made on metachromatic elements constitute a truly enormous literature. The reviews of MICHELS and NAGEL, cited below, are examples of the extent of this literature; each review is concerned with only one class of chromotrope. What is called for, then, is a survey of critical observations and a presentation of the patterns into which metachromatic elements fall. Chromotropes *in situ* are first described with respect to distribution and localization, merely touching for the sake of continuity their assumed roles and compositions. Chromotropes *in vitro* are then treated in detail, with special emphasis on those compounds presumed to be responsible for metachromasy in cells and tissues.

I. Chromotropes *in situ*

A. Plants

1. S c h i z o p h y t a

Numerous bacteria were early found to possess cytoplasmic granules or vacuoles that were metachromatic. NAGEL (1948) assigns the discovery of the bacterial bodies to ERNST and to NEISSER in 1888 and to BABES in 1889. BABES (1895) felt that his "metachromatischen Körperchen" were nuclear analogs and claims the first description of them in the tuberculosis and leprosy bacilli and in the fungus of actinomycosis, noting that similar bodies occurred in yeasts. According to NAGEL, the names for these metachromatic or basophilic granules were numerous: Babes-Ernst granules, metachromatic granules, metachromatin, Neisser's bodies, polar granules. red granules, sporogenic granules. Figures of these elements are shown by BABES (1895) and DOBELL (1908).

GRIMME, in 1902, described vacuoles or globules in bacteria which he called "Volutanskugeln," a name based on his methods for displaying similar globules in *Spirillum volutans,* where they were especially numerous (NAGEL 1948). MEYER (1904) gave this substance the name *volutin,* finding it also in molds, yeasts and fungi. MEYER's definition of volutin was based chiefly on his sulfuric acid-methylene blue staining method, still used today.

NAGEL's (1948) review should be consulted for further references on volutin and related bodies described in bacteria and other forms. She presents a table of the principle studies of volutin, discusses earlier work, and records the divergent opinions on chemical properties and functions assigned to volutin. In a summarizing statement, NAGEL says that "differences of opinion concerning the morphology, physiology and chemical nature of (volutin) are so great that it is at present impossible to equate results."

Although many earlier investigators related volutin to nuclear components or to nucleic acid, SMITH, WILKINSON and DUGUID (1954) find evidence to the contrary. In *Aerobacter aerogenes*, only a metaphosphate fraction showed complete parallelism with staining of volutin. BELOZERSKY (1945, 1947) believes that the volutin of *Spirillum volutans* is an unusual complex, a nucleotide containing sulfuric acid in ester linkage.

The cell walls or capsules of some bacteria are readily stained by a number of older methods employing metachromatic dyes. The nature of bacterial cell surface layers is discussed by DUBOS (1949). Claims of cellulose, hemicellulose and chitin are not substantiated in the membranes and walls; the slime layers and capsules contain various polysaccharides, proteins and lipids. The polysaccharides, especially the acidic ones, might then be the extracellular metachromatic or basophilic components of bacteria. Hyaluronic acid obtained from pneumococci and group A streptococci (MEYER 1938; PIKE 1950) is undoubtedly a capsular polysaccharide. A polysaccharide from *Serratia marcescens* (SHEAR and TURNER 1943) is metachromatic in solution (HEILBRUNN and WILSON 1950).

2. Thallophyta

Yeasts are best known for *intracellular* metachromatic elements, notably volutin. While volutin is found in many algae, fungi and molds, only in the yeasts has there been a concerted effort to identify the chromotrope. In algae, the abundant slimes and mucilages are *extracellular* chromotropes.

General descriptions of volutin in these plants are available (GUILLIERMOND, MANGENOT, and PLANTEFOL 1933; HENRICI 1930; VAN HERWERDEN 1917; KLEIN 1929; LINDEGREN 1945; MEYER 1904; NAGEL 1948). Through all the earlier work, there runs a dominant theme relating volutin to nuclear components if not to nucleic acids themselves. This was the conclusion of CASPERSSON and BRANDT (1941), who studied yeast volutin by means of the ultraviolet microspectrophotometer. Volutin, for these authors. contains RNA and is equivalent to the nucleolus and heterochromatin of higher plants. *Metachromatin,* coexisting with volutin, may be a polysaccharide sulfuric ester. In *Saccharomyces cerevisiae,* WIAME (1946. 1947 a, 1947 b, 1947 c, 1949) presented evidence that volutin, responsible for the metachromasy of these yeast cells, was distinct from the basophilic nucleic acid. Volutin was identified as *metaphosphate* by WIAME. NAGEL (1948) summarizes the features of the two entities in controversy:

(1) A metachromatic vacuolar (or cytoplasmic) substance has been called *volutin, metachromatin, metaphosphate.*

(2) A ribonucleic acid-containing substance in the cytoplasm has been called *volutin, basophilic granules.*

Metachromasy has been reported in a myxomycete, *Didinium nigripes*. in cytoplasmic granules distinct from other vacuoles (SKUPIENSKI 1929).

From numerous marine algae, especially *Rhodophyceae* and *Phaeophyceae*, many substances of medicinal or commercial value are extracted. Little is known about the localization of these extracts; they probably arise in the outer cell wall and certainly form part of the slippery, mucilaginous layers (WHISTLER and SMART 1953). Except for algin, all the seaweed polysaccharides are notable for their high sulfuric acid content (MORI 1953). Some of them are among the strongest chromotropes, whose metachromasy in solution is clearly established (Table 5).

3. Spermatophyta

No reports of metachromasy inside the cells of higher plants have been found. Extracellular metachromasy, as in the lower plants, is better known by studies of extracts in solution than by any histological or cytological localization. The flowering plants yield extracts which are often ill-defined gums and gels; these are collectively the *polyuronides* or polymers of uronic acid. They include the gums, slimes and pectinaceous materials, many of which are metachromatic (Table 5). Consult WHISTLER and SMART (1953) for information on the origin and chemical properties of the plant polysaccharides.

Even though the distribution of chromotropes among plants is heavily weighted in favor of the lower plants, only part of this may be due to the large numbers of studies relating to those forms. There are also chemical reasons for thinking that the distribution of metachromatic staining has a significant basis. It seems permissible to summarize the distribution and approximate nature of plant chromotropes as follows:

(1) Metachromasy in lower plants is both extracellular and intracellular, while that of higher plants is exclusively extracellular.

(2) The intracellular chromotropes of lower plants are characterized by their phosphate-content (nucleic acid, metaphosphate), although sulfate has been mentioned as a possibility.

(3) The extracellular chromotropes of lower plants are characterized by sulfate-content, only rarely by uronic acid.

(4) The extracellular chromotropes of higher plants are characterized by uronic acid-content.

B. Animals

1. Protozoa

The trypanosomes contain intensely basophilic or metachromatic granules, typically in the flagellate half of the cell, which have been called *chromatoid granules* implying nuclear material (LAVERAN and MESNIL 1901; MINCHIN 1909; WOODCOCK 1906) or *volutin* (BERGHE 1947; SWELLENGREBEL 1908). The relation of this material to yeast volutin is not entirely clear, no chemical studies having been made on the protozoan variety comparable to those on *Saccharomyces*. However, VAN DEN BERGHE believes the volutin of Protozoa represents the site of accumulated ribonucleic acid (based essentially on ribonuclease digestion), though there is no certainty that the granules are entirely composed of that substance. WOODCOCK observed that an unfavorable environment increased the number of chromatoid granules in a cell, in contrast to yeast volutin which increases in actively-metabolizing cells in a phosphorus-rich medium (HERWERDEN 1917; WIAME 1947 a).

VAN DEN BERGHE (1947) also finds volutin in the Rhizopoda, Sporozoa and Ciliophora. The occurence of volutin in those forms capable of ingesting food materials requires proof that the "volutin" does not represent bacteria or algae (NAGEL 1948).

MAST and DOYLE (1935 a, 1935 b) found certain bodies in *Amoeba proteus* with an outer layer containing "metachromatin" and a non-starch carbohydrate. MAST and DOYLE consider metachromatin, based on a staining reaction, to be nucleic acid plus an unknown base.

Metachromasy in the ciliates is not common. MÖLLENDORFF (1924) cites a report of metachromasy in *Paramecium* following vital staining with rhodamine O. SESACHAR (1950, 1953) observed nuclear metachromasy in *Chilodenella uncinatus*. Certain granules in the macronucleus were metachromatic with toluidine blue, red with Giemsa, Feulgen-negative, and green with fast green. SESACHAR refers to a similar pattern in *Chilodon cucullulus* earlier described by REICHENOW. The evidence of SESACHAR points to ribonucleoprotein as the substance responsible for metachromasy in the nucleus, though it does not rule out other possibilities.

A single observation was made some time ago on the trichocysts of *Paramecium caudatum,* extruded directly into a dilute solution of toluidine blue (KELLY, unpublished observation). The trichocysts were seen to be metachromatic (red). No great importance is attached to this observation, since it is difficult to reliably assign color to objects as thin as trichocysts, especially when viewed in an aqueous medium. However, the "trichocyst might ... consist of tectin, a pseudochitin similar in composition to mucin," with the trichocyst membrane protein resembling collagen (JAKUS 1945).

2. Echinodermata

E. B. HARVEY (1941) reported metachromasy in the hyaline layer (after centrifugation) and in the jelly of *Arbacia punctulata* eggs. LANSING and ROSENTHAL (1949) found that the slight metachromasy in the cortex or vitelline membrane of this egg was enhanced after ribonuclease digestion, while basophilia in the surface region was abolished by the digestion. The possibility of a polysaccharide sulfuric ester at the surface of the egg was suggested by LANSING and ROSENTHAL. In *Arbacia*, toluidine blue metachromasy was found in the egg jelly of ovarian eggs and shed eggs, in the cortical region of the egg, and in the ovarian ground substance (KELLY 1954). The metachromasy persisted in egg jelly which had been dialyzed against distilled water.

The egg of the starfish, *Asterias forbesi,* is only slightly metachromatic in the cortex and jelly (KELLY 1954). There is also metachromasy in the ovarian connective tissue ground substance. Cytoplasmic metachromasy of the *Asterias* egg is variable, and never alcohol-resistant.

The voluminous egg jelly of the sand dollar, *Echinarachnius parma,* is intensely metachromatic, a reaction which is strongly resistant to alcohol (KELLY 1954). Faint metachromasy is also seen in the cortex or vitelline membrane of the egg. The ovary is practically non-metachromatic. Fig. 4 shows absorption spectra of the nucleolus, cytoplasm and jelly of the *Echinarachnius* egg after staining with toluidine blue.

To what may the metachromasy of echinoderms be assigned in terms of chemical components? While little is known about the connective tissue ground substance of these forms, it is likely that metachromasy in the ovarian connective tissue reflects the presence of acid mucopolysaccharides. The nature of the egg jellies is better known. VASSEUR (1948 b) reports a high content of sulfuric acid (about 26% sulfate) in the jelly of the sea urchin, *Strongylocentrotus droebachiensis* Müll. Fucoidin, a seaweed polysaccharide sulfate (MORI 1953), is found in sea urchin egg-jelly (VASSEUR 1948 a). MONNÉ and HÅRDE

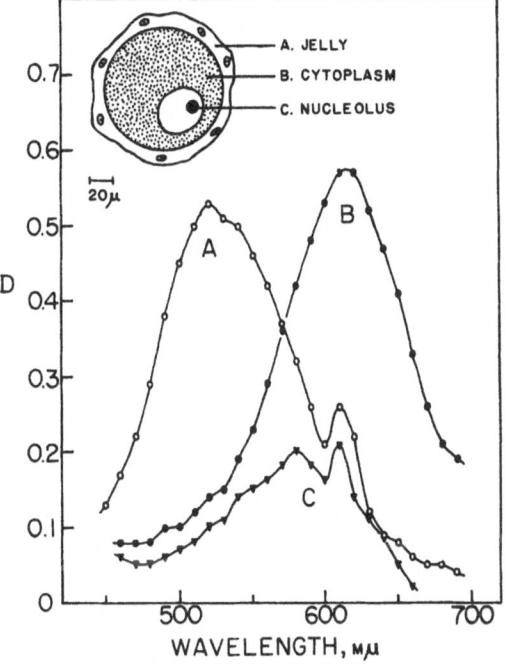

Fig. 4. Regional absorbance in the mature ovarian egg of *Echinarachnius parma,* the sand dollar, sectioned at 7 microns and stained with toluidine blue. The area measured in each case is the same as that of the nucleolus, about 95 μ². The resulting spectra represent the intensely metachromatic *red* of the jelly (*A*), *blue* of the cytoplasm, and *violet* of the nucleolus. (In curves *A* and *C*, secondary peaks at about 610 mμ undoubtedly represent dye fixed in the orthochromatic form.)

(1951) claim that sea urchin egg cortical granules contain a polysaccharide sulfuric ester. The physiological implications of such compounds will later be discussed.

3. Annelida

Vital staining with toluidine blue has revealed a metachromatic substance in the egg of the marine tube-worm, *Chaetopterus pergamentaceus* (KELLY 1950, 1954). The metachromatic vacuoles or granules move centripetally with light centrifugation, they have not been observed before germinal vesicle breakdown, and they are not resistant to alcohol. Except for unidentified metachromatic masses, the *Chaetopterus* ovary is non-metachromatic. It would be highly desirable to know the nature of the metachromatic material in *Chaetopterus* eggs, since it has been implicated

in a theory of "cellular homeostasis" (GOLDSTEIN 1953).   Cortical metachromasy in *Nereis limbata* eggs, assignable to the granules which are the
precursors of the jelly extruded upon fertilization, is described by KELLY
(1954).   Cortical and jelly metachromasy in this form are extremely
resistant to alcohol. Variable metachromasy was also found in the nucleo-
plasm and on the nucleolus of this egg.   The only other intracellular
chromotropes known to be in annelid eggs are the cortical granules of the
egg of a polychaete worm, *Tylorrhincus heterochaetus* (YAMAMOTO 1952).

No metachromasy was found in the eggs of *Arenicola marina* or *Am-
phitrite brunnea* (KELLY 1954).

FERRY (1939) says that the jelly of *Arenicola* eggs is essentially a polysaccharide. The jelly of *Nereis limbata* eggs is apparently a uronic acid-
containing carbohydrate (COSTELLO 1949). It is interesting that the tube in
which *Chaetopterus variopedatus* dwells is made of concentric lamellae of
mucoprotein (EWER and HANSON 1945).

4. M o l l u s c a

Metachromasy in the eggs and ovaries of various clams, *Mytilus edulis,
Spisula solidissima* and *Venus mercenaria,* is described by KELLY (1954).
Various degrees of metachromasy were found in the jelly, cortex, cytoplasm
and nucleus of eggs and in the ovarian tissue. THOMAS (1951, 1954) has
investigated *Spisula solidissima,* the surf clam, and has extracted heparin-
like anticoagulants from the eggs and from various tissues of this form.
These anticoagulants are related to the various metachromatic elements;
they are present in the mucous secretions and in the connective tissue
ground substance.   Histochemical, chemical and clinical studies of the
Spisula metachromatic substances have also been made by FROMMHAGEN,
FAHRENBACH, BROCKMAN and STOKSTAD (1953) and LOVE and FROMMHAGEN
(1953).   According to these authors, the anticoagulant ("mactin-A") does
not arise from mast cells and probably consists of four different acid
polysaccharides.

These substances may be related to the process of calcification in
molluscs (BEVELANDER and BENZER 1948; BEVELANDER 1952).   ALLEN (1951 b)
thinks that a mucopolysaccharide is transferred from a nucleolar structure
(nucleolinus) to the spindle during cleavage of the *Spisula* egg. An anticoagulant has been extracted from the mucus of *Charonia lampas* (see
THOMAS 1954); it is a sulfuric acid ester. EWER and HANSON (1945) feel that
the differences in certain staining reactions (including those of some metachromatic dyes) of molluscs and other invertebrates is due to differences in
mucoproteins.

5. A r t h r o p o d a

Data are fragmentary concerning metachromasy in these animals.
ANDERSON (1950) describes the metachromasy associated with certain glands
in the male reproductive tract of the Japanese beetle, *Popillia japonica*
Newman. Mast cells (basophils?) are said to be found in insect blood

(MICHELS 1938). Metachromatic inclusions, probably yolk substances, are found in the oocytes or fertilized eggs of the prawn, *Palaemonetes vulgaris,* and fiddler crab, *Uca pugnax* (KELLY 1954).

6. Tunicata

The test-cells associated with the egg of an ascidian, *Cynthia roretzi* Drasche, contain a mucin-like substance (HIRAI 1949). In *Ciona intestinalis* eggs, although the test cells are negative, the chorion and certain bodies ("foam bodies") closely applied to the chorion are metachromatic (KELLY 1954).

7. Pisces

According to MICHELS (1938), fishes are the lowest animals whose tissues contain in large amounts the best-known of all metachromatic elements—the mast cell of EHRLICH. Beyond noting the fact that mast cells are found in fishes and all higher phyla, discussion of these cells will be deferred to the section on Mammalia, where a number of chromotropes are classified into characteristic groups.

Metachromasy and other histochemical tests of cortical alveoli in the eggs of two salmons, *Onchorhyncus keta* and *O. nerka,* led KUSA (1954) to conclude that the alveoli are "composed of mucoprotein containing poly-saccharide esters with sulfuric acid" and they "may also contain amino sugar." The chorion and ovarian cells associated with eggs of the toadfish, *Opsanus tau,* were stained metachromatically red to violet with toluidine blue (KELLY 1954).

8. Amphibia

MICHELS (1938) says that the relation between blood basophils and tissue mast cells is closer in lower vertebrates than in higher ones; the distinction between the two is entirely lacking in amphibians and lower forms.

RE (1951) studied jelly of eggs and oviducts in a toad, *Bufo vulgaris,* and a frog, *Rana esculenta,* using metachromasy among other tests to follow changes in the jelly with ultraviolet irradiation. Although both types of jelly lost metachromasy with prolonged exposure to UV, differences in the two types of jelly were found regarding viscosity changes after irradiation at different wavelengths and after exposure to hyaluronidase. KELLY (1954) describes differences between the jelly deposited on normal *Rana pipiens* eggs and the jelly deposited on "glass eggs." Vital staining with toluidine blue leads to metachromasy in the inner layer of jelly on normal eggs, none in the outer layers. When glass beads pick up jelly in normal fashion by passage through the oviducts, no metachromasy is found in any jelly layer. The suggestion is made that the "jelly substance responsible for metachromasy may come from the egg and not from the oviduct." This finding may have a bearing on the fact that many amphibian eggs cannot be fertilized until they have accumulated jelly by passage through the upper third or so of the oviduct.

9. Reptilia, Aves

The tissues and blood of the turtle have long been known as good sources of mast cells and an exceptionally high count is found in horned toads and lizards (MICHELS 1938). On the other hand, MICHELS says that tissue mast cells are sparse in birds and the blood basophils are numerous. BERDNIKOW and CHAMPY (1932) isolated a mucoid from the rooster's comb which was metachromatic with thionine. Instead of sulfate, however, it contained 1.5% phosphorus.

10. Mammalia

This section will be somewhat less than a full consideration of mammalian chromotropes since it will not list the large numbers of observations that have been made on metachromasy in mammalian tissues. It will, in another sense, be more than a discussion of mammalian chromotropes because patterns of metachromasy perceived in the mammalian tissues can so often be extrapolated to the lower forms. Simply knowing where metachromasy occurs is not the question. If that were so, we would be in the position of the histologist prior to LISON's time, accumulating more and more information about a phenomenon whose meaning is obscure. Not only LISON's discovery of the meaning of histological metachromasy but also the development of chemical techniques for handling the variety of chromotropes encountered have added a new dimension to a classical method. While it is true that chemical studies virtually opened up the metachromatic reaction, there are a few cases where metachromasy has led the way toward solution of a stagnant biochemical problem. Recognizing its faults, there can be no question that the metachromatic reaction has secured a firm position in the battery of standard histochemical tests.

The mammalian chromotropes are discussed under four categories: *intracellular*, exemplified by the mast cell, *extracellular*, being chiefly the mesenchymal ground substance, *secretions* of epithelial origin, and *pathological* "substances" or conditions. In each case, it is important to note events associated with the appearance or depression of metachromasy. Such changes may be long-term in connection with growth (HOLMGREN 1949) and aging (LOEWI 1953), they may be short-term as in the rapid degranulation of mast cells at a site of injury (LARSSON and SYLVÉN 1948), or they may be cyclical (FELL 1953, discussion by DEMPSEY). The general distribution of mammalian chromotropes is surveyed in the papers of BUNTING (1950) and WISLOCKI, BUNTING, and DEMPSEY (1947 b).

a) Intracellular chromotropes

The *mast cell* resides in the connective tissue, its cytoplasmic granules being intensely metachromatic under virtually all conditions. It is practically a rule that the distribution of mast cells closely parallels that of loose connective tissue. Consequently, the parenchyma of organs tends to be low in mast cells while organ capsules or trabeculae may contain large numbers. Mast cells are found in all mammals to a greater or lesser

degree and they are present in many lower animals. Excellent reviews
are available on the mast cell. LEHNER (1924), MICHELS (1938) and NEU-
MANN (1932) cover exhaustively the earlier literature. ASBOE-HANSEN (1954 b),
FRIBERG, GRAF, and ÅBERG (1951), and PETTERSON (1954) emphasize more
recent investigations on the mast cell, especially those relating to its
functions, its histochemistry and its changes under irradiation. Table 3 is

Table 3. *Distribution of Mast Cells.*

(Based chiefly on MICHELS, 1938.)

Site or Condition	Few	Moderate	Many
Organism	Bird Guinea pig Rabbit	Cat Dog Man Monkey	Fishes Horned lizard Mouse Rat
Organs, tissues	Adrenals Bone Cartilage Epithelium Kidney Liver	Lymph nodes Ovary Pancreas Salivary glands Testis	Bladder Lung Mammary gland Mesenteries Prostate Tongue Uterus
Connective tissues	Old Hyaline Sclerotic		Young Loose Finely fibrillar
Pathological States	Acute inflammation Scar tissue Tumor stroma		Chron. inflammation Granulation tissue Tumor periphery

a summary of the distribution of mast cells, taken especially from the
review of MICHELS. It is only possible to suggest the patterns of mast
cell distribution in such a table; the details are found in the reviews cited
above.

It is a widely-held opinion today that the mast cell is the source of
the potent anticoagulant, *heparin.* Nothing played a larger part in this
identification than did the metachromatic reaction. This role of a simple
staining procedure in a larger chemical investigation is highly instructive.
The main features of the "heparin story" are well-told by JORPES (1946).

In 1916, McLEAN, a student in HOWELL's laboratory, discovered a liver
phosphatide fraction with thromboplastic activity. When the substance was
freed of cephalin, however, it had a marked anticoagulant action. By 1918,
HOWELL and his co-workers had given this "antiprothrombin" the name *heparin.*
They still considered it a phosphatide. Subsequently, JORPES and his group in
Sweden systematically attacked the chemistry of heparin and identified the

latter as a *mucoitin sulfuric acid*. One of the keys to the problem was the high ash content, which had been recorded but ignored by the Howell group. At the time of Jorpes' work, Best, Charles, Scott and others in Toronto were helping to establish the chemical nature of heparin, especially by means of highly purified samples. Jorpes then discovered that heparin was strongly metachromatic in solution, a discovery which led to the identification by Holmgreen and Wilander (1937) of Ehrlich's mast cell as the site of heparin in the tissues. The main points in the identification are (1) the intense metachromasy of heparin and mast cell granules, (2) a correlation between tissues containing numerous mast cells and tissues yielding large amounts of heparin, and (3) the circumstantial occurrence of large numbers of mast cells along smaller blood vessels.

Unquestionably, the older evidence for assigning heparin to the mast cell is good and confirmation for this idea has since come from several types of investigation. In none of the earlier extractions was the cell-population characterized except that tissues were selected to be relatively high or low in mast cells. Under a buffer extraction method gentler than most heparin extractions, ox-liver capsules containing 40% mast cells yielded a heparin-lipoprotein complex which had greater anticoagulant activity than a purified heparin (Snellman, Sylvén and Julén 1951; Sylvén 1951). It was assumed that this complex resembles the native state of heparin. Oliver, Bloom and Mangieri (1947) were able to isolate fifty times as much heparin from a dog mast cell tumor as could be isolated from the richest normal source, although the anticoagulant potency of their purified preparation was low. Paff and Bloom (1949) cultured tumor mast cells from the dog and observed the release of a metachromatic substance, presumed to be heparin, into the culture medium. Clotting tests of this substance were negative. Heparin has been isolated from a scrotal tumor containing numerous mast cells (126 mg. "purified" heparin or 11.1 g. crude heparin from 10 kg. tumor tissue in a scrotum weighing 22 kg.) (Ehrich, Seiffer, Alburn, and Begany 1949). This is the only known report of heparin from a human source. Riley (1953) has also correlated heparin, histamine and mast cells, based on his studies of human and dog mast cell tumors and ox pleura.

Certain questions remain unanswered. Has it been proven that heparin, as we know it today, is obtainable from a single cell type? Only extractions of pure cultures will answer this question. How can heparins be extracted from tissues that contain no mast cells, such as clam tissues (Thomas 1951, 1954; Frommhagen, Fahrenbach, Brockman, and Stokstad 1953; Love and Frommhagen 1953)? Why has no heparin been extracted from the fishes, said to contain the greatest numbers of mast cells (Michels 1938)? Is it sufficient to say that a tissue has "numerous" mast cells or has a "high content", when it is in fact difficult to specify cell-populations in metazoan tissues (Stowell 1952; Pettersson 1954)? Compton (1952), in an extensive study of the hamster mast cell, concludes that "although evidence indicates that mast cell granules contain heparin, this content is still unproven" and a "survey of the distribution of mast cells within the hamster does not support their supposed function as 'heparinocytes'.'' Regarding the latter conclusion, however, it must be pointed out that no one has attempted to extract heparin from the hamster, although Koksal (1953) obtained a heparin-like substance from the mouse connective tissue, which was assumed to come from the mast cell.

There are differences of opinion on the cytological localization of the metachromatic material in the mast cytoplasm, most authors believing that the granules are metachromatic (Friberg, Graf, and Åberg 1951; Montagna, Eisen, and Goldman 1954) and others that it is the intergranular matrix which is metachromatic (Sylvén 1951). Another role is suggested for the mast cell by Asboe-Hansen. He says that "Ehrlich's mast cells, which secrete hyaluronic acid, are interpreted as the peripheral transmitters of hormonal action on the connective tissues" (Asboe-Hansen 1950) and that this secretion of the mesenchymal polysaccharide occurs "perhaps by way of a sulphuric preliminary stage resembling heparin" (Asboe-Hansen 1951). Kelsall and Crabb (1954) think that mast cells may provide mucopolysaccharides for the secretion of mucus. These are not new ideas (Michels 1938); it is felt that they are worthy of serious investigation but that they contain inherent difficulties of proof greater than those associated with proof of the heparin content of mast cells.

The metachromatic component of mast cells is not the only element of interest. It should be mentioned that a number of enzymes, lipids and ribonucleic acid have been localized in the mast cell cytoplasm (Montagna, Eisen, and Goldman 1954) and histamine may be a significant product of the cell (Ehrich 1953; Fawcett 1954; Riley 1953).

Next to the mast cell, the blood *basophil* has attracted most attention for its intense basophilia or metachromasy. While the two cells seem closely related, metachromasy may be their only common feature (Michels 1938). The literature on basophils is reviewed by Michels (1938), Neumann (1932) and Speirs (1955). Virtually nothing is known about these cells. Their striking metachromasy led Tötterman (1948) to investigate the possibility that they might contain heparin. He did this indirectly by seeking a parallel between basophilia (metachromasy) and bleeding tendency in myeloid leukemia. Not finding such a correlation, it was concluded that basophils are not "heparinocytes." Basophils attract attention because their number is not great in the blood and because they are so intensely basophilic. But it is tedious to work with cells requiring large counts, no matter how easily they are seen in blood smears or counting chambers. The basophil has thus been rather slighted by working hematologists. It is to be hoped that our knowledge of the basophil will be increased following the recent development of a simple and ingenious method for counting these cells, based on the use of a metachromatic dye (Moore and James 1953).

Perhaps it is of historical interest only that both Michels (1938) and Neumann (1932) suggest the association of volutin in plant cells with the cytoplasmic granules of both mast cells and basophils.

3*

The cytoplasmic metachromasy of other cells is of an entirely different order from that of mast cell and basophil. "Faint cytoplasmic metachromasia is observable under various conditions in many cells of the body" but is ordinarily abolished by ribonuclease digestion (WISLOCKI, BUNTING, and DEMPSEY 1947 b). DUSTIN (1947) thinks that the metachromatic granules or vacuoles seen in erythrocytes and reticulocytes is due largely to the concentration of dye in these structures, that the distinction between orthochromatic and metachromatic staining in these cells may represent a difference in the physical state of ribonucleoprotein. GROSSFELD (1954) observed metachromasy in the fibroblast and other cells when vitally stained with toluidine blue, which was attributed primarily to concentration of dye by the living cells. A somewhat specific staining of the Golgi zone of fibroblasts was attained in fibroblasts by using an azure B fraction extracted from methylene blue (LASFARGUES and DI FINE 1950). Metachromasy is found within certain alveolar cells of the lung (BERTALANFFY and LEBLOND 1953). The β-cells of the hypophysis, as well as the colloidal contents of follicles in the pars intermedia, are metachromatic (BIENWALD 1939; HERLANT 1943; ROMEIS 1940).

Delicate metachromasy is found in the cytoplasm of nerve cells and glial cells (WISLOCKI, BUNTING, and DEMPSEY 1947 b), presumably due to ribonucleic acid. It is also seen in Schwann cell cytoplasm in degeneration and regeneration, in the Schwann cell "protagon granules of Reich," and in the normal myelin sheath (NOBACK 1954). NOBACK thinks the metachromasy of the Schwann cytoplasm is due to ribonucleoprotein, that of the protagon granules to a phosphatide or cerebroside, and that of the myelin sheath to a phosphatide or sulfatide. He says that chondroitin sulfate is eliminated as a possible chromotrope in the Schwann cytoplasm but that mucoitin sulfate is not. The metachromasy of myelin has been reported by others (CHANG 1938; FEYRTER 1936; FEYRTER and PISCHINGER 1942). Metachromasy in brain cells and myelin is cautiously related to phosphate by LANDSMEER (1951), while WISLOCKI and SINGER (1950) think that myelin metachromasy may indicate sulfatides. KRÜCKE (1939) considers certain pathological states where nerves exhibit metachromasy.

Nuclear metachromasy is clearly shown by the colored figures of VON MÖLLENDORFF (1924). It is ordinarily not seen in sections which have been exposed to alcohol. CARNES, WEISSMAN, and RUBIN (1951) find chromatin and chromosomes metachromatic when isolated liver nuclei are stained and examined in the hydrated state. This is related to the metachromasy of desoxyribonucleic acid in solution (WEISSMAN, CARNES, RUBIN, and FISHER 1952). Chromosomes digested with trypsin leave metachromatic residues; this metachromasy and the Feulgen reaction are abolished, in turn, by nuclease (YASUZUMI, MORI, MATSUKURA, and MINAMINO 1950). It is to be expected that the low order of metachromasy displayed by nuclear components would indicate nucleic acid. However, ALLEN (1951 b) has implicated nucleolar mucopolysaccharide in mitotic activity of the spindle, SHARP (1943) makes the undocumented statement that a "sulfuric acid ester" is found in the nucleolus and MONNÉ and SLAUTTERBACK (1950) find nuclear mucopolysaccharides in the sea urchin egg. On the general subject of nuclear metachromasy, it has been recognized for a long time that the hue (s) of nuclear metachromasy is not the same as that of the mast cell or cartilage. Thus, VON MÖLLENDORFF used the term "semi-metachromatic" to describe nuclear metachromasy, while LISON prefers not to use the term in relation to differences in hue but applies it to those substances which are metachromatic only under certain conditions (LISON and MUTSAARS 1950).

b) Connective tissue ground substance

There is the possibility of finding metachromasy wherever the mesenchyme or its adult derivatives occur. In practice, there is a rather constant distribution of metachromatic elements in the mammalian con-

Table 4. *Metachromasy in Mammalian Ground Substance.*

Structure	Localization Within Structure	References
Arteries	Intima (small a.), entire wall (elastic a.)	3, 8, 12
Bone	Matrix prior to calcification	4, 10
Cartilage	Matrix	3, 6, 12
Connective tissue, fetal . .	Matrix	12
Eye	Cornea, sclera, retina, vitreous	3, 9, 11, 12
Heart	Valve rings	3
Joints, synovial	Synovial fluid, cartilage	12
Mammary gland	Perilobular stroma	3, 12
Nasal passages	Mucosal matrix	12
Nucleus pulposus	Matrix	12
Ovary	Zona pellucida, follicular fluid, coronal ground substance	12
Prostate	Perilobular stroma	3
Skin	Papillary dermis, hair follicles, "sex-skin"	1, 3, 12
Teeth	Predentin, dentin, stellate reticulum	5, 10, 14
Thymus	Hassall's corpuscles	12
Trachea, bronchi	Mucosal matrix, tunica propria, cartilage	3, 12
Umbilical cord	Wharton's jelly, umbilical vessels	3, 8
Uterus	Placenta, decidua, endometrial stroma	2, 7, 12, 13

[1] ARGYRIS 1954; [2] BENSLEY 1934; [3] BUNTING 1950; [4] FELL 1953; [5] LEBLOND, BÉLANGER and GREULICH 1955; [6] LOEWI 1953; [7] MCKAY 1950; [8] ROMANINI 1951; [9] SIDMAN and WISLOCKI 1954; [10] SOGNNAES 1955; [11] WISLOCKI 1952; [12] WISLOCKI, BUNTING and DEMPSEY 1947 b; [13] WISLOCKI and DEMPSEY 1948; [14] WISLOCKI and SOGNNAES 1950.

nective tissue. The matrix of cartilage is comparable to the mast cell cytoplasm in the degree of metachromasy shown. Almost as metachromatic as cartilage are the matrix or ground substance of the cornea, Whartons jelly of the umbilical cord and the nucleus pulposus of intervertebral discs. General descriptions of normal connective tissue and its ground substance, including observations on metachromasy, are found in histology textbooks and in the papers of Angevine (1951), Gersh (1952), Lillie (1952 a, 1952 c), and Wislocki, Bunting, and Dempsey (1947 b). A number of papers, other than those specifically cited here, have appeared in the five conference reports on *Connective Tissues* of the Josiah Macy Jr. Foundation individually cited in this paper and in a conference report on a symposium held under the auspices of the New York Academy of Sciences (Durans-Reynals 1950).

In order to conserve space, Table 4 was compiled to show the general distribution of metachromasy in the ground substance.

The compounds likely to be responsible for metachromasy in the ground substance are *mucopolysaccharides* or *mucoproteins* (Meyer 1938, 1947, 1953). A mucopolysaccharide is a hexosamine-containing compound, which may or may not contain hexuronic acid. Those that contain uronic acid are the *acid* mucopolysaccharides; they may also contain sulfuric acid in ester-linkage. The *neutral* mucopolysaccharides contain hexoses instead of uronic acid.

A number of older terms describe tissue products that exhibit degrees of metachromasy. While these terms have no chemical significance, it is true that they describe what the eye sees under various conditions of staining and it is only within this framework that they should be used. Perhaps the most general term is *mucin*, referring to any viscid secretion in the animal body. *Mucus* is that mucin secreted onto a free surface by epithelial cells, while *mucoid* is the mucin secreted by cells of the mesoderm. Any mucin presumably contains mucopolysaccharides or mucoproteins. These older terms, as Meyer (1938) suggests, should be used only in a physiological sense. *Pseudomucin* represents an "atypical" mucin, possibly an "aged" mucin, commonly associated with cysts (e. g., ovarian cystadenomas). *Mucinoid* is also an atypical mucin, sometimes used in the same sense as *mucoid*. However, it should be noted that *mucoid* has also a more restricted definition, meaning a carbohydrate in covalent union with protein, such as serum mucoid. Meyer does not believe these "true mucoids" are found in connective tissues, although others feel that they may be there (Ragan 1953, pp. 16—46). In any case, such true mucoids would hardly be metachromatic and would require localization by other means.

There is no question of *in vitro* metachromasy of chondroitin sulfate. This mucopolysaccharide is the metachromatic component of cartilage, a tissue which may contain in the neighborhood of twenty per cent chondroitin sulfate (Strandberg 1950). Einbinder and Schubert (1950) concluded that there is "no basis for believing that salt-like compounds can exist at neutral reaction between mucopolysaccharides and collagen or that cartilage could be such a salt," in general agreement with Partridge but opposed to the view of Meyer and his co-workers, who considered cartilage to be a protein salt of chondroitin sulfate. But Meyer (1951, 1953) has repeatedly emphasized that the nature of the protein with which acid mucopolysaccharides are combined in tissues is unknown, that the protein is probably *not* collagen itself, that the protein is, in fact, characterized by a high content of aromatic amino acids.

There is less agreement on the metachromasy of hyaluronic acid, some thinking that hyaluronate is metachromatic in tissue sections, smears and solutions (Braden 1955; Bunting 1950; Schlechter and Campani 1948; Wislocki, Bunting, and Dempsey 1947 b; Levine and Schubert 1952 b) while others question this reaction (Meyer 1947; Sylvén and Malmgren 1952). Sylvén and Malmgren doubt that the concentration of hyaluronate in tissues is ever high enough to yield a visible metachromatic reaction, feeling that the contribution of hyaluronic acid to the metachromasy of the ground substance is actually unknown.

The formulas below are presented for convenience only. They are modified from Whistler and Smart (1953), who indicate the tentative nature of such structural proposals.

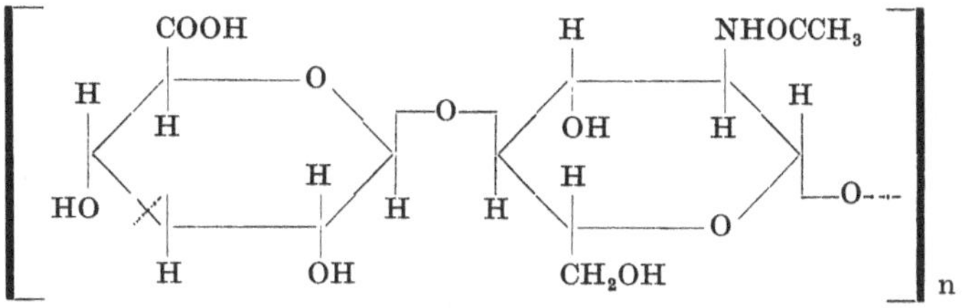

Hyaluronic acid

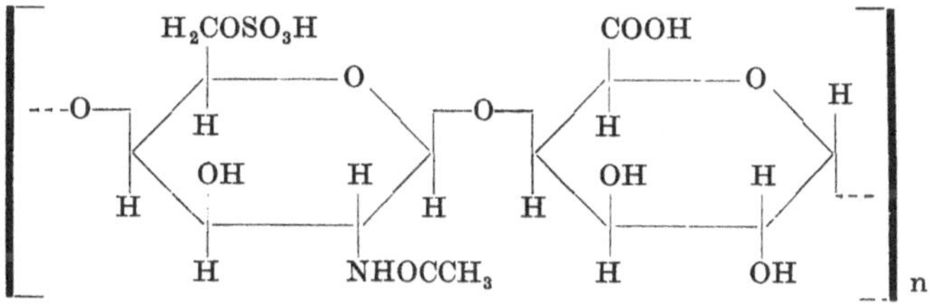

Chondroitin monosulfuric acid

Chondroitin sulfate and hyaluronic acid are not the only mucopolysaccharides with which the histochemist may have to reckon. Meyer and Rapport (1951) describe three chondroitin sulfates, for example, which are distinguished not only by distribution but also by properties (precipitability, optical rotation) which elude the present histochemical techniques. Newly-discovered mucopolysaccharides, *keratosulfate* (Meyer, Linker, Davidson, and Weissman 1953) and *chondroitin* (Davidson and Meyer 1954) from bovine cornea, are among such substances. While keratosulfate may be the only sulfated polysaccharide of animal origin not containing uronic acid, there is evidence that polysaccharides may exist in the connective tissues which are not acidic at all (Consden and Bird 1954; Glegg, Eidinger, and Leblond 1954). In the work of Glegg and co-workers, it is interesting to note that those fractions containing no uronic acid displayed a carbohydrate pattern similar to that obtained in tissues containing much reticulin and basement membranes.

The relation of sulfated polysaccharides to calcification processes, normal and abnormal, is one of the most interesting involvements of the metachromatic reaction. This is a particularly direct relation in the case of the hard tissues, which were the subject of a recent conference (GREEP 1955). The histochemistry and histogenesis of cartilage, bone, cementum, dentin and enamel are discussed in particular by SOGNNAES (1955) and by LEBLOND, BÉLANGER, and GREULICH (1955). Two contrasting opinions seem to be held on the role of sulfated compounds (metachromatic components) in calcification. LEBLOND and his co-workers, as well as SOBEL (1955), visualize sulfated compound as local factors in calcified areas, these compounds serving to bind the mineral components to the protein matrix. SOGNNAES suggests that the sulfated acid mucopolysaccharides, instead of forming a link between the fibrous and crystalline elements, may actually maintain the *uncalcified* state and provide metabolic pathways through relatively acellular and avascular regions. The avascularity of metachromatic areas has been noted elsewhere (BUNTING 1950). It is instructive to compare normal calcification involving sulfated polysaccharides with abnormal calcification, as in arteries, where the deposition of cholesterol also tends to occur at sites of metachromasy (FABER 1949).

The origin of mesenchymal polysaccharides is of the greatest interest, though practically nothing beyond speculation has been proposed. The fibroblast (GERSH, in RAGAN 1953 a) and the mast cell (ASBOE-HANSEN 1954 a, 1954 b) are the cells currently nominated as sources of mucopolysaccharides. The distinction may ultimately lose meaning if "regeneration of mast cells begins ... in cells indistinguishable from fibroblasts in the walls of small blood vessels" (FAWCETT 1953).

c) Secretions

Closely related to the mesenchymal mucins, are those mucins secreted onto a free surface by epithelial cells of ectodermal or entodermal origin. *Mucus* varies widely in its reaction to stains which are supposed to be "specific" for the various secretions. LILLIE (1949 b) made an exhaustive study of normal and pathological mucins in man and other mammals. It is not possible to summarize his observations in brief fashion but the "plurality of the mucins seems strongly indicated." LILLIE's thirty-four detailed tables should be consulted by the interested reader.

Ordinarily, there are no glands secreting metachromatic materials onto the skin surface. Only the sweat glands might be expected to secrete such substances and no metachromasy was associated with the cells or secretions of human eccrine or apocrine sweat glands (BUNTING, WISLOCKI, and DEMPSEY 1948). The glands secreting onto moist surfaces, however regularly contain and produce metachromatic substances. These glands include the major and minor salivary glands as well as glands of the respiratory, intestinal and reproductive tracts (LILLIE 1949 b; WISLOCKI, BUNTING, and DEMPSEY 1947 a, 1947 b).

The variability of staining reactions, including metachromasy, is well known among the mucous secretions. This may be a complex variability, involving concentrations of metachromatic components and the suppressing effect of non-metachromatic components, as well as fundamental chemical differences in the chromotropes. Thus, while acidic mucopolysaccharides are found in mucus generally, it is well-known that gastric mucin also contains a neutral fraction (MEYER 1938). The acid mucopolysaccharide of mucus is *mucoitin sulfuric acid* which contains equimolar portions of glucosamine, glucuronic acid, sulfuric acid and acetic acid. It thus differs in its amino sugar from chondroitin sulfate and is more closely related to heparin.

FELL (1953), in a paper largely concerned with the effects of vitamin A on bone growth, discusses some interesting aspects of secreting epithelia. There is a general relation between vitamin A deficiency and keratinization of epithelia, while an excess of the vitamin has the opposite effect, even inducing mucous metaplasia of (chick) skin. DEMPSEY, in the discussion of FELL's paper, noted that some mucosae (e. g., in the vagina) alternate cyclically between keratinization and mucus-production. It was even remarked that excess vitamin A can lead to the production of cilia in an ectodermal epithelium which is ordinarily not ciliated. These observations are related to the different forms in which sulfur is found in epithelial structures: keratinous epithelia are characterized by sulfhydryl (—SH) or disulfide (—SS—) sulfur, while the secretions of non-keratinized epithelia are sulfuric esters (—OSO$_3$—). Unqualified assumptions must not be made on the interconversions of these two types of sulfur compounds (MEYER, in FELL 1953).

d) Pathological chromotropes

The title of this section is not meant to imply that there are necessarily pathological "substances" which stain metachromatically. While some metachromatic elements familiar to the pathologist may prove to be quite distinct from ordinary tissue components, there is a general impression that these elements represent a more or less *altered* connective tissue. The connective tissue and its ground substance are treated from the pathologist's standpoint by ALTSCHULER and ANGEVINE (1951), ASBOE-HANSEN (1950), GERSH and CATCHPOLE (1949), and KLEMPERER (1950). Many pathological deposits of the type to be discussed are recognized on the basis of gross appearance, physical consistency, natural color, staining reactions, general distribution in the tissues, and relation to the pattern of a disease state.

Amyloid is a colorless, firm, waxy or glassy, intercellular material which infiltrates a number of organs, the kidney, liver and spleen in particular. It is commonly eosinophilic, takes up Congo red in a selective though unexplained manner, and stains with iodine in a fashion resembling starch. The metachromasy of amyloid is peculiar and variable. It does not stain metachromatically with any but the triphenylmethane dyes, like methyl violet or crystal violet. CARNES and FORKER (1954) confirm this "unexplained superiority of triphenylmethanes" as amyloid stains when compared with toluidine blue. Toluidine blue, on the other hand, is metachromatic with cartilage while crystal violet is not. No difference in response to crystal violet was found between "primary" and "secondary" amyloid. CARNES and FORKER cannot identify amyloid with chondroitin sulfate, an old idea that appears in a number of textbooks on pathology. VEEN and MEIJER (1948) analyzed a para-amyloid tumor and concluded that amyloid was not a chondroitin sulfate, that perhaps both substances are separately deposited and coexist in a diseased organ. This is only one of several factors that might explain the variability of amyloid staining. KRAMER and WINDRUM (1955) emphasize the impurity of dyes as a factor which may lead to *differential* staining of amyloid rather than to the "so-called metachromasia" (see also p. 23).

Perhaps amyloid, instead of being "starch-like," resembles cellulose more closely. One of the early reactions of amyloid was a *blue-black* color with iodine after sulfuric acid. The uptake of Congo red by amyloid is also well-known. In the direct dyeing of cotton fabric by Congo red, there may be a correct "fit" of dye and substrate (BRODE 1955) mediated through hydrogen bonds (SINGER 1952). Does amyloid represent a re-routing of carbohydrate metabolism, whereby the usual animal structural polysaccharides are sacrificed in favor of plant-like polysaccharides resembling celluloses or hemicelluloses?

(After BRODE 1955)

Fibrinoid is a homogeneous, acellular, eosinophilic, refractile substance which, according to BENNETT (1951), was named by NEUMANN in 1880. In contrast to the title of his paper, BENNETT prefers the phrase "fibrinoid change" to "fibrinoid degeneration." He reviews some of the thoughts on fibrinoid pathogenesis: occurrence at hematoparenchymal barriers, imbibition of plasma protein, exudation and precipitation of fibrin, "retrograde process" in the connective tissue, or precipitation of acid mucopolysaccharide in the ground substance. The morphologic sequence of events in fibrinoid change is: interfibrillar material increases, metachromasy may appear or increase, fibrils become swollen and refractile and may fragment, acidophilia succeeds metachromasy, amorphous granular débris may appear, with eventual liquefaction. BENNETT thinks that metachromasy may signal an initial stage in fibrinoid change, while ANGEVINE says the metachromatic material is laid down before fibrinoid changes begin. ALTSCHULER and ANGEVINE (1949) note a crystal violet metachromasy of fibrinoid, somewhat akin to that obtained with amyloid. ANGEVINE says that he has never seen fibrinoid develop in an area that did not have metachromasy

associated with it (Meyer 1951). Apparently, fibrin and fibrinoid are similar
in their general staining properties, both resembling a plasma clot (Singer
1954; Singer and Wislocki 1948). Overstaining apparently leads to metachromasy of these substances. In a general discussion, Dempsey remarked
on the *inhibition* of metachromasy by fibrin (Bennett 1951).

Bunting (1950) outlines and discusses the distribution of metachromasy
in tumors other than mast cell tumors, Aschoff bodies, and in newly-
formed granulation tissue, as well as the metachromasy of normal tissue.
Various mucin-producing tumors were investigated by Grishman (1952),
Hempelmann (1940), Lennox, Pearse, and Richards (1952), and Lillie (1949 b),
yielding certain common observations. The differences in tumor mucins
with respect to metachromasy are generally in intensity of staining rather
than specific differences. However, Hempelmann claims the ability to
distinguish mesenchymal and epithelial mucins (mucoproteins) based on the
dilution of metachromatic dyes. In mixed tumors, the cartilaginous areas
resemble normal cartilage and may be distinguished from mucins within
acini. No normal animal tissue was comparable to myxoma tissue. Perhaps
the most general observation is that tumor mucin resembles the normal
mucin secreted by the tissue of origin.

II. Chromotropes *in vitro*

Lison's (1935 a, 1935 b, 1936 a) investigation of a wide variety of chemical
compounds clearly showed that metachromasy could be produced in solution. Choosing at random a large number of alcohols and phenols,
ketones, esters and organic acids, lipids. carbohydrates and proteins,
Lison tested these compounds in a solution of metachromatic dye (0.005%
toluidine blue). His major conclusions concerning chromotropes were:

(1) All metachromatic substances are sulfuric esters of high molecular
weight.

(2) If the sulfuric radical is removed, chromotropic properties are lost.

(3) If the sulfuric radical is introduced into a non-metachromatic compound of high molecular weight, a chromotrope is formed.

Thus, starch, cellulose and glycogen were not metachromatic but their
sulfuric half-esters were. Chondroitin sulfate from cartilage was found to
be an excellent chromotrope, while chondroitin was not metachromatic.
Lison did not find gums, mucilages, hemicelluloses and pectic materials to
be chromotropes, although a few years later some of these plant products
were re-investigated elsewhere and found to be metachromatic. Nitric and
phosphoric esters (including nucleic acid) and one silicic ester (lichenin
from *Citraria islandica,* the Iceland moss) were not metachromatic according to Lison. It is now known that nucleic acids and inorganic silicates
are metachromatic.

The importance of Lison's discovery is illustrated by some curious facts. The
mast cell and the matrix of cartilage were known for many years to be the most
metachromatic of all histological elements. According to Strandberg (1950), the

Table 5. *Chromotropes in vitro.*

Some degree of metachromasy has been shown by these natural, substituted or synthetic compounds, in sols, gels or films. Descriptive remarks are those of the original author or are based on common usage. Metachromasy with cationic dyes is implied except for those chromotropes marked with an asterisk (*), where significant color changes were observed with anionic dyes.

Parent Substance	Description	References
Sulfuric esters, sulfonates :		
Agar	Galactose polymer, algae, *Gelidium*	1, 15, 16, 17, 22, 23, 24, 30
Alginic acid	Polymannuronate, algae, "Paritol"	31, 37
"Anticoagulant" . . .	Extract, surf clam, *Spisula*	34, 35
Arabic acid	Gum, *Acacia*	16, 17
Carageenan	Extract, Irish moss, *Chondrus*	1, 11, 16, 17
Cellulose	Plant cell walls	16, 17, 27
Chitin	Polyglucosamine, crustacean shells	16, 37
β-Cholestanol		8
Cholesterol		8, 16, 17
Chondroitin sulfate . .	Ground substance, connective tissue	1, 3, 13, 14, 15, 16, 17, 28, 30, 31, 37
Dextran	Glucose polymer	25, 31, 37
Dextran	Carboxymethyl-, glucose polymer	37
Geloses	Extracts, algae (6 genera)	16
Glucose		16, 17
Glycogen		16
Heparin(s)	Mast cell, other cells (?)	3, 11, 12, 13, 15, 29, 31, 37
Hyaluronic acid . . .	Ground substance, connective tissue	31
Inulin	Fructose polymer, bacterial	31
Levulan	Fructose polymer, *Dahlia*	31
Mactin-A	Extract, clam, *Mactra*	7, 19
Mactin-B	Extract, clam, *Artica*	7
Mucoitin sulfate . . .	Mucus, epithelial "mucin"	11, 16
Naphthol		16, 17
Pectin	Polygalacturonate, fruit, "Treburon"	3
Polysaccharide	Galactose:arabinose polymer, larch	31
Polyvinyl sulfate . . .		33
Saccharose		16, 17
Starch		16, 17
Thymol		16, 17
Xylan	Pentosan, wood, "Thrombocid" "T. S. 144"	37
Carboxylates		
Alginate	Polymannuronate, algae, "Kelacid"	15, 31, 33
Aerosols, #18, #22 .	Long-chain hydrocarbons, sulfonated	22
Arabinate	Gum, *Acacia*	1
Cellulose	Carboxymethyl-	22, 31

Parent Substance	Description	References
Carboxylates		
Dextran	Carboxymethyl-, glucosan, wood	31, 37
Hyaluronic acid . . .	Ground substance, connective tissue	3, 15, 27, 28, 31, 32, 37
Myristic acid		15
Oleate		22
Pectate	Polygalacturonate, fruits	1
Pectinate	Polygalacturonate, fruits	1, 31
"Slime"	Flaxseed, *Linus*	1
Stearate		33
Phosphates		
Desoxyribonucleate . .	(Metachromatic within certain limits)	1, 18, 22, 23, 31, 38, 39, 40
Dextran		31
Lecithin	Egg	1, 33
Metaphosphate	Molds, yeasts ("volutin")	6, 31, 39, 40, 41
Ribonucleate	(Metachromatic within certain limits)	1, 18, 22, 23, 31, 38
Phosphatides	(Weak metachromasy)	1, 10
Inorganic compounds		
Bases, strong	(Possibility of dye base formation)	1, 16
Salts	(Usually *inhibit* metachromasy)	1, 16, 22, 40
Silicates	Quartz particles, soluble silicates	5, 20, 21
Miscellaneous		
Detergents, anionic . .		4, 15, 31
*Detergents, cationic . .		4, 15
Fatty acids, fats . . .	(Weak metachromasy)	10
Gelatin	(Not metachromatic when pure)	1
*Guanidine		1
Jelly, egg-	Echinoderms (8 genera)	13, 36
Mucopolysaccharides .	Cornea, dentin, gastric mucin	3
*Novocaine		1
Polysaccharide	Bacterial, *Serratia marcescens*	9
*Proteins	Blood serum	26
Pseudomucoid	Rooster's comb	2
*Quinine		1
*Strychnine		1

[1] BANK and BUNGENBERG DE JONG 1939; [2] BERDNIKOW and CHAMPY 1932; [3] BRADEN 1954; [4] CORRIN and HARKINS 1947; [5] CURRAN 1953; [6] DAMLE and KRISHNAN 1954; [7] FROMMHAGEN, FAHRENBACH, BROCKMAN and STOKSTAD 1953; [8] GOLDENBERG and GOLDENBERG 1955; [9] HEILBRUNN and WILSON 1950; [10] HOLTFRETER 1946; [11] JAQUES, BRUCE-MITFORD and RICKER 1947; [12] JORPES 1946; [13] KELLY 1955; [14] LEVINE and SCHUBERT 1952 a; [15] LEVINE and SCHUBERT 1952 b; [16] LISON 1935 a; [17] LISON 1935 b; [18] LISON and MUTSAARS 1950; [19] LOVE and FROMMHAGEN 1953; [20] MERRILL and SPENCER 1948; [21] MERRILL, SPENCER and GETTY 1948; [22] MICHAELIS 1947; [23] MICHAELIS 1950; [24] MICHAELIS and GRANICK 1945; [25] MOWRY 1954; [26] RAWSON 1943; [27] SCHLECHTER and CAMPANI 1948; [28] SCHUBERT and LEVINE 1953; [29] SNELLMAN, JENSEN and SYLVÉN 1949; [30] SPEK 1944; [31] SYLVÉN 1954; [32] SYLVÉN and MALMGREN 1952; [33] TERUYAMA 1954; [34] THOMAS 1951; [35] THOMAS 1954; [36] VASSEUR 1948; [37] WALTON and RICKETTS 1954; [38] WEISSMANN, CARNES, RUBIN and FISHER 1952; [39] WIAME 1947 a; [40] WIAME 1947 b; [41] WIAME 1949.

chemistry of chondroitin sulfate from cartilage was known as early as 1854 when BOEDECKER extracted a "Chondroitsäure." SCHMIEDEBERG renamed this "chondroitin sulfuric acid" in 1891 and a tetrasaccharide structure was proposed in 1925 by LEVINE. Not until 1935 was the conspicuous metachromatic component of cartilage directly identified with chondroitin sulfate by LISON. On the other hand, there was available no chemical substance of any kind, bearing the slightest relation to the mast cell, until 1916 when McLEAN began his studies on the liver fraction which was later called heparin. Even then, it was not until *after* LISON's work that JORPES discovered the metachromasy of heparin and HOLMGREN and WILANDER, in 1937, concluded that the metachromatic granules of the mast cell cytoplasm were heparin.

The next significant step was taken by BANK and BUNGENBERG DE JONG (1939), who systematically investigated a smaller number of chromotropes than did LISON. They found that sulfuric esters were not the only chromotropes: colloids bearing carboxyl and phosphoryl groups were also metachromatic. The main conclusions of BANK and BUNGENBERG DE JONG were:

(1) The minimum requirement for metachromasy is opposite charge of dye and substrate.

(2) Metachromatic change increases with charge density of a compound, that is, metachromasy is inversely related to equivalent weight of chromotrope.

(3) The effectiveness of anionic groups responsible for the charge of a colloid is in the order, sulfate > phosphate > carboxyl.

BANK and BUNGENBERG DE JONG also studied the influence of inorganic salts, acid, base and alcohols on the metachromatic reaction. They were the first to suggest that there might be a metachromatic reaction of acidic dyes with cationic colloids.

Table 5 is a summary of compounds which, to a greater or lesser degree, induce metachromasy in appropriate dyes. Most of these chromotropes require no comment, a number of them having been mentioned already. Certain features of chromotropes in general are more appropriately discussed in the next section.

The Reaction

Numerous features of the reaction between a metachromatic dye and a chromotrope have been anticipated in earlier sections. It has been shown, for example, that a solution of dye alone displays characteristics facilitating the prediction of a metachromatic reaction with a chromotrope. Some of the chromotropes listed in Table 5 are metachromatic under virtually all conditions and others are only metachromatic within specified limits. Both groups may be distinguished from substances which never display metachromatic activity under any circumstances. It is the purpose of this section to review the array of metachromatic reactions and the influence of various factors upon them. While the main emphasis is placed on objective observation, the associated visual appearances are of equal importance and are described in each case.

A. General Characteristics

The most direct description of metachromasy is found in a family of absorption curves showing the spectral shifts which occur when a chromotrope is added to a dilute solution of metachromatic dye (Fig. 5). Many authors have published similar curves for different dyes and chromotropes (JAQUES, BRUCE-MITFORD, and RICKER 1947; KELLY 1955 b; LEVINE and SCHUBERT 1952 a; LISON 1953; MICHAELIS and GRANICK 1945; MICHAELIS 1947, 1950; SYLVÉN 1954; SYLVÉN and MALMGREN 1952; WALTON and RICKETTS 1954). As a first approximation, there is general agreement on the decrease of the α-peak and a simultaneous development of the γ-peak at a *lower* wavelength. The position of the γ-peak depends to a slight degree on the concentration of the dye (MICHAELIS and GRANICK (1945).

It is clear that, regardless of visual observations, the absorption shifts may be recorded for a large number of dyes. It is sometimes true that the spectrophotometer will show striking absorption changes for a dye which is practically never used for metachromatic staining. This is the case for methylene blue, whose spectral changes are particularly sharp (MICHAELIS 1944), but which is inferior to azure A or toluidine blue in histological practice.

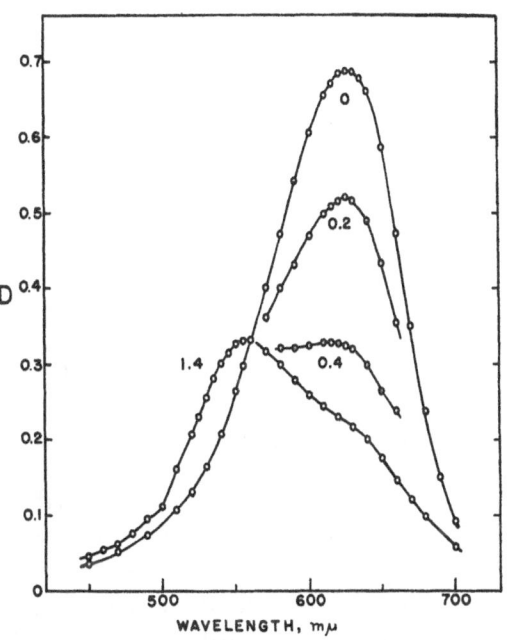

Fig. 5. Spectra of toluidine blue in the presence of chondroitin sulfate. Dye concentration, 1.63×10^{-5} M. Chondroitin sulfate concentrations (mg%) are shown by numbers in the figure. The curves representing chondroitin sulfate concentrations of 0.2 mg% and 0.4 mg% are abridged for clarity. A concentration of 1.4 mg% of this chromotrope induces virtually maximum metachromasy. (Coleman spectrophotometer, model 14.)

Superimposed upon any basic metachromatic shift in wavelength, an additional requirement for *visual* metachromasy is that the shift must take place near the wavelength of maximum relative luminosity, at about $555\,m\mu$ (p. 10). This is the reason for the selection of "good" metachromatic dyes whose spectral shifts are less or no better than dyes which are not used for metachromatic staining (Table 2). On this basis, a number of dyes are metachromatic in a fundamental sense even though the color changes are not striking. SCHUBERT and LEVINE (1953) discuss several of these dyes whose metachromasy has been controversial.

B. Variations in the General Pattern

LEHNER (1924) remarks on the variety of shades or hues that are seen in stained sections with many dyes, not only with the accepted metachromatic dyes. How has it come about that, even before the general use of spectrophotometers, metachromasy was recognized as a phenomenon somewhat apart from the occurrence of slight changes in hue in a stained section? The answer is completely lacking in rigor, depending on the reliability of the human eye and upon the experience of the observer. There is enough controversy concerning the "simple" objective observations without entering into sterile argument on such matters of opinion. But lest the human be underestimated, who ultimately decides what is objective and what is subjective, the remark of no less an observer than Brachet is quoted: "As regards the sensitivity of the trained human eye, there is no doubt that there are cases where gradients can be picked up much more easily with the eye than with other methods. In the particular case of Amphibian eggs, quantitative measurements of the U. V. absorption would be especially difficult because of the great heterogeneity of the cells, which contain yolk, pigment, etc. However, such gradients are very obvious on mere examination of stained preparations" (Pollister and Ris 1947, discussion by Brachet, p. 155).

Bank and Bungenberg de Jong (1939) defined metachromasy on the basis of visual *and* spectroscopic observations. Visually, there is a change in *color tone* (hue) and a change in color *intensity* (brilliance). Spectroscopically, there is the appearance of new absorption bands and an intensity change or loss of bands characteristic of the aqueous solution. Thus, virtually any color change would technically be metachromatic. Taking toluidine blue as an example, the orthochromatic *blue* is changed toward *violet* or *red* (by addition of chromotrope or base) or toward *green* (in the presence of acids, salts, excess nucleic acid).

The first case to be discussed is that of relatively minor shifts of absorption maxima toward shorter wavelengths, that is, in the same sense as the metachromasy of heparin and toluidine blue. This involves the important question of the metachromasy of nucleic acids. Some authors (Michaelis 1947, 1950; Wiame, 1947 b; Lison, 1935 a) believe that nucleic acids are not metachromatic or that they even inhibit metachromasy. Others (Carnes, Weissman, and Rubin 1951; Flax and Himes 1952; Himes and Flax 1950; Weissman, Carnes, Rubin, and Fisher 1952; Wislocki, Bunting, and Dempsey 1947 b) observe metachromasy of nucleic acids in solution or of nucleic-acid containing structures in histological preparations. The distinction between nucleic acid metachromasy and that of a chromotrope like heparin is one of degree and stability. Whereas, in toluidine blue solutions, heparin induces maximal metachromasy with a shift in absorption maximum from 630 mμ to about 540 mμ, both DNA and RNA (Weissman, Carnes, Rubin, and Fisher 1952) similarly lead to "absorption maximum shifts to the γ-region" but the new peaks arise at only 560–570 mμ. The metachromasy of sulfated compounds is generally *red* while that of nucleic acids is *violet*. In addition, the conditions under which nucleic acid metachromasy occurs are more circumscribed than those of sulfated compounds. Weissman and his co-workers find that pentose and desoxypentose nucleic acids induce metachromasy in toluidine blue between pH 6 and 7, at temperatures below 30° C., when the ionic strength of the solution is less than 0.03, and when the molar ratio of dye to nucleic acid phosporus is between 0.4 and 1.4. Few authors, incidentally, have so carefully set the parameters for their observations of metachromasy. The general shift of absorption maxima toward shorter wavelengths has been called *positive* metachromasy, regardless of magnitude of the shift. A positive metachromatic shift which occurs to a limited degree and only under certain conditions, as in the case of nucleic acids, is called *semi-metachromatic* (Lison and Mutsaars 1950).

Opposite to the usual metachromatic shifts, there occurs a shift of absorption maxima toward *longer* wavelengths where the color deepens. For a dye like toluidine blue, this would mean *greenish* hues. Such shifts are usually of a low order of magnitude. They lead to a state of *negative* metachromasy (Lison and Mutsaars 1950). This has been observed in the epidermis (Lehner 1924), nuclei or nucleic acid solutions (Bank and Bungenberg de Jong 1939; Lison and Mutsaars 1950), and even in mast cell cytoplasm (Holmgren and Wohlfart 1948). It is related to an excess of chromotrope (Lison and Mutsaars 1950; Michaelis 1947) and also resembles the colors and spectra observed when metachromatic dyes are dissolved in organic solvents (Lehner 1924; Michaelis 1947).

There are certain observations which constitute a miscellany of metachromasy in the sense that they have not been clearly distinguished from the usual metachromatic reactions or have not yet been studied sufficiently to permit any conclusions. Kelley and Miller (1935 b) mention a color shift of pararosanilin which resembles the difference in color between an aqueous and an alcoholic solution of the dye, equivalent to "protein metachromasy." They also observed a "hematoxylin metachromasy" which was apparently of an entirely different order than that obtained with thionine, being largely dependent on pH (Kelley and Miller 1935 a). There is the interesting possibility of metachromasy of anionic dyes (Bank and Bungenberg de Jong 1939), whose importance for the general theory of metachromasy is emphasized by Levine and Schubert (1952 a). It is questionable whether the relatively minor shifts of absorption maxima of anionic dyes (Rawson 1943; Corrin and Harkins 1947) are to be explained on the same basis as "positive" metachromatic shifts. They resemble more the effects observed when basic dyes are influenced by high temperature, acids or organic solvents. In Rawson's studies, for example, she found that the peaks of the anionic dyes were shifted by plasma albumin about 20 mμ toward *longer* wavelengths in every case, suggesting that there occurs "a dampening of the bond-energies of the dye as binding with protein takes place." Changes in hue and intensity of dyes adsorbed to proteins are usually slight, in any case, and hardly of the order associated with most metachromatic reactions.

The question of "false" metachromasy has arisen, especially in connection with histological staining. Lison (1935 a) defined the specificity of the metachromatic reaction for sulfuric esters in terms of those conditions where only the normal form of the dye was stable—low pH, presence of alcohol or other non-aqueous media, and relatively low dye concentration. BANK and BUNGENBERG DE JONG (1939), even while producing evidence that metachromasy is induced by non-sulfated compounds, endorsed the legitimacy of LISON's "réaction d'ester sulfurique" in view of the conditions under which it must be carried out in order to constitute a histochemical reaction. SYLVÉN (1941, 1945) originally drew rather sharp lines between the "true" metachromasy of sulfuric esters and the "false" metachromasy of other elements. Later (SYLVÉN and MALGREN 1952), it was said "we require the appearance of a new absorption band gamma (γ) at a shorter wavelength than the 'normal' alpha (α) and/or beta (β) band" and "we also require a certain stability of the new aggregate compounds (responsible for the γ-band)." It is true that only the appearance of γ-bands underlies striking visual changes and that anything less tends to lose visual significance. At the present time, there is little point in making any distinctions on the basis of hue; a series of compounds could be selected to induce a nearly continuous array of hues from blue to red in a dye like azure A. It is the relative stability of metachromatic complexes that SYLVÉN (1954) currently emphasizes.

Even if the shifts in absorption of some dyes are not particularly striking to the eye, it is extremely important to investigate them for theoretical reasons. The relative changes in α-, β-, and γ-bands with increasing concentration of chromotrope, for example, are most readily apparent when they are plotted in the manner of LEVINE and SCHUBERT (1952 a, 1952 b; SCHUBERT and LEVINE 1953). Reference to this work will later be made in several connections.

C. Influence of the Chromotrope

1. Chromotrope structure

The general nature of chromotropes is apparent from Table 5 and the minimum requirements were outlined on page 46. Other than the fact that metachromasy increases with charge density (BANK and BUNGENBERG DE JONG 1939), there is very little known about the influence of chromotrope structure on metachromasy. Possibly an intercharge distance on the chromotrope of about 5 A is required (for azure A) and the degree of metachromasy increases as this distance decreases (SYLVÉN 1954). Steric hindrance undoubtedly occurs in many chromotropes (and in dyes) to modify the relatively simple effects of dissociation of ionized groups and their spacings (SHEPPARD and GEDDES 1944 a; SYLVÉN 1954). Although the spectral shifts associated with concentration (MICHAELIS 1950) of metachromatic dyes or with production of metachromasy by chromotropes (SYLVÉN 1954; WEISSMAN, CARNES, RUBIN and FISHER 1952) represent increase in bond-energies of the order of 3–12 kcal./mole, this tells nothing about the nature of the bonds. There is no real explanation, in the absence of structural information, for the magnitudes of metachromatic shifts induced by various chromotropes. In toluidine blue, the order of magnitude of these shifts is as follows:

cholesterol sulfate, 150 mμ (GOLDENBERG and GOLDENBERG 1955); heparin, 90 mμ; chondroitin sulfate, 75 mμ; nucleic acids, 50—60 mμ (WEISSMAN, CARNES, RUBIN, and FISHER 1952). SYLVÉN (1954) presents a table illustrating this for azure A. (See also Table 2, page 12.)

While it has been assumed that a relatively high molecular weight is characteristic of chromotropes (LISON 1935 a), this may not be so. WALTON and RICKETTS (1954) studied a number of sulfated glucose polymers, chiefly dextrans, and concluded that extent of polymerization was not important above a limit of about 4 glucose units. Between 4 and 6000 units, metachromasy was related only to sulfur content. It was further calculated that one molecule of toluidine blue is bound by each anionic group of the substrates.

2. Chromotrope concentration

For a given concentration of metachromatic dye, maximum metachromasy is obtained with equivalent amounts of different chromotropes, although the magnitude of the maximal metachromasy depends on the nature of the chromotrope (BANK and BUNGENBERG DE JONG 1939). That the "magnitude" of metachromasy is not a simple expression is shown when the absorption at the wavelengths of the α-, β-, and γ-peaks are plotted against chromotrope concentration (LEVINE and SCHUBERT 1952 a, 1952 b; SCHUBERT and LEVINE 1953). The main observations of these authors are as follows:

(1) Initially, α- and β-peaks fall with the addition of chromotrope, while the γ-peak is rising.

(2) As the chromotrope concentration reaches that of the dye, absorption of all bands levels off and then declines with further addition of chromotrope.

LEVINE and SCHUBERT's figures should be consulted for details of important variations within this general pattern. Chondroitin sulfate and hyaluronate were added to methylene blue and to crystal violet: in all four cases there were significant differences in reaction. In addition these authors have studied dye bound in relation to metachromasy by means of equilibrium dialysis (LEVINE and SCHUBERT 1952 b) and specific conductance (SCHUBERT and LEVINE 1953). Their general conclusions from this work will be reserved for discussion on the theories of metachromasy.

WIAME 1947 b) defined metachromasy as $\mathfrak{E}_{530}/\mathfrak{E}_{630}$ for toluidine blue and used this measure to compare the metachromasy of metaphosphate with other chromotropes. Similar ratios have been used by other authors for solution studies (ALLEN 1951 a; KELLY 1955 b; LEVINE and SCHUBERT 1952 a; HAMERMAN and SCHUBERT 1953) and for microspectrophotometry of tissue sections (FLAX and HIMES 1952). These ratios are particularly convenient in making certain comparisons among chromotropes, providing the more complex spectral changes of the dye are known.

The suppression of metachromasy in the presence of a large excess of chromotrope is the general rule (BANK and BUNGENBERG DE JONG 1939; LEVINE

and SCHUBERT 1952 a; LISON and MUTSAARS 1950; SYLVÉN 1954; WIAME 1947 b).
MICHAELIS (1947) distinguishes several basic modes of interaction between
dye and substrate as concentration of the latter is varied. The initially
orthochromatic dye solution becomes metachromatic as the chromotrope is
added until the concentration of the latter approaches that of the dye. In
this range of concentration coagulation or flocculation of the chromotrope
may occur. Further addition of chromotrope leads to maximal metachromasy which may or may not then be suppressed by a large excess of
chromotrope. MICHAELIS thought, at the time, that the suppression was a
"depolymerizing" effect of the chromotrope whose anionic groups were
increasingly free to exert electric repulsion and maintain a colloidal
solution. Certain chromotropes (carboxymethyl cellulose, agar) were,
however, metachromatic under all conditions and this was only partially
explained as an effect of large numbers of hydrophilic (hydroxyl) groups,
in addition to the anionic groups. MICHAELIS (1950) later questioned the
polymerization theory, in view of the staining of agar gels, where metachromasy occurs with such an excess of agar that it seems unlikely that
the dye is adsorbed in polymolecular form.

The occurrence of precipitates between chromotropes and dyes is
somewhat variable, although many authors have observed them (CORRIN
and HARKINS 1947; MÖLLENDORFF 1924; SPEK 1940; SYLVÉN 1954; SYLVÉN and
MALMGREN 1952). SPEK (1944) proposed a special name—"Fällungsmetachromasie"—for the occurrence of these complexes. LEVINE and SCHUBERT
(1952 a) have this to say about precipitates: while they occur at sufficiently high concentration of chromotrope, there is no trace of them when
concentrations of dye and chromotrope are approximately equal and
when minimum extinctions of α- and β-peaks occur. This is somewhat
below the concentrations of chromotrope where maximum metachromasy
occurs. WALTON and RICKETTS (1954) find that "with high concentrations
of strongly-charged acidic polysaccharides the metachromatic complex precipitates from solution." SYLVÉN (1954) studied heparin-azure A solutions
before and after centrifugation at 40,000 g. for an hour. In all solutions
precipitates were formed, although the heparin concentration was always
greater than the dye concentration. BANK and BUNGENBERG DE JONG (1939),
however, observed that a *maximally* metachromatic gum arabic-toluidine
blue solution formed no precipitate in four days.

D. Influence of Other Agents

1. Solvents

Common organic solvents suppress both the dye-chromotrope reaction
and the absorption shift occurring with concentration of dye alone (LEHNER
1924; LISON 1935 a). The effectiveness of alcohols in suppressing the metachromasy of a sodium arabinate-toluidine blue solution is in the order,
n-butanol $>$ n-propanol $>$ ethanol $>$ methanol, according to BANK and
BUNGENBERG DE JONG (1939). EDLUND and PERSSON (1949) find that the amount
of a given solvent required to abolish the metachromasy of any concen-

tration of a given chromotrope is relatively constant. Different chromotropes require different amounts of the solvent. SYLVÉN (1954) says that the metachromasy of moderately-charged chromotropes withstands ethanol up to about 30%, while the metachromasy of some ester sulfates and sulfonates is stable up to 50–70% ethanol. SNELLMAN, JENSEN, and SYLVÉN (1949) find that anticoagulant and metachromatic activities of heparin are separate molecular functions, based on solvent studies.

The relative stability of metachromasy in the presence of organic solvents was the most practical basis for LISON's (1935 a) original specifications for the histochemical employment of the reaction to detect sulfuric esters, endorsed by SYLVÉN (1941, 1945).

2. Salts. Ionic strength

LISON (1935 a) briefly states that salts suppress metachromasy and that barium chloride is more effective in this respect than sodium chloride, although the salt effect was not studied extensively. BANK and BUNGENBERG DE JONG (1939) investigated a large number of salts and found that they *suppressed* the metachromasy induced by a chromotrope and were themselves metachromatic in higher concentrations. For colloids and cationic dyes, the salt effect increases with valence of the inorganic cation, the inorganic anion being relatively unimportant. The ability of inorganic salts to cause metachromasy is presented in a table by BANK and BUNGENBERG DE JONG. No simple statement can be made about inorganic salts as chromotropes, except that those anions which ordinarily form polyvalent complexes seem to be better chromotropes.

The metachromasy of metaphosphate is well-known and is related to yeast volutin (WIAME 1946, 1947 b, 1947 c, 1949). WIAME (1947 b), SCHMIDT (1951), and DAMLE and KRISHNAN (1954) describe the use of metachromasy in quantitative studies of metaphosphate. WIAME found that hexametaphosphate was intensely metachromatic, trimetaphosphate only slightly so, ortho- and pyrophosphate and adenosine triphosphate were non-metachromatic, while nucleic acid (type unspecified) and potassium sulfate inhibited metachromasy. SYLVÉN (1954) also studied a series of phosphates and found that the metachromasy of greatest stability was obtained with potassium metaphosphates of relatively large particle size, including a metaphosphate from *Aspergillus niger.*

SYLVÉN (1954) remarks on the tendency of inorganic anions to suppress the aggregation of dye molecules and, at higher ionic strength, on the salting-out effect which leads to *orthochromatic* precipitates. He and others (BANK and BUNGENBERG DE JONG 1939; SPEK 1940) note exceptions to this behavior for iodide and thiocyanate, which are unexplainedly metachromatic. LEVINE and SCHUBERT (1952 a) find that salts at a concentration of about 1 M produce a metachromatic effect. The increase of ionic strength suppresses metachromasy (WEISSMAN, CARNES, RUBIN, and FISHER 1952) and has a marked effect on the shapes of absorption curves (SYLVÉN and MALMGREN 1952).

3. pH

Low pH suppresses metachromasy and higher pH enhances it (LEHNER 1924; LISON 1935 a; BANK and BUNGENBERG DE JONG 1939; WEISSMAN, CARNES, RUBIN, and FISHER 1952). At the higher pH levels, there is the possibility of formation of dye base and many of the bases have colors similar to the metachromatic color (LEHNER 1924; LISON 1935 a). In some cases, this may only occur above pH 11 (SCHLECHTER and CAMPANI 1948), although WEISSMAN. CARNES, RUBIN, and FISHER (1952) report a *decline* of metachromasy in the neighborhood of pH 8. LISON performed several experiments showing that metachromasy did not depend on a substance, presumably the dye base (HANSEN 1908), which could be extracted from the aqueous dye solution by an organic solvent. First, a solution of dye is still capable of staining metachromatically when several successive extractions have removed all of the material in question. Second, LISON found that metachromasy exists at a pH where significant formation of dye bases is unlikely. The relatively stable metachromasy of sulfuric esters, in fact, exists at pH 3 or lower. Finally, if a solution of brilliant cresyl blue is alkalized, it develops a red-orange color, presumably due to the formation of dye base at pH 10–12. A similar color is developed in a second tube of brilliant cresyl blue by the addition of chomotrope. Both tubes are shaken with chloroform. From the alkaline solution, a red-orange color passes into the chloroform phase. From the solution containing the chromotrope, nothing passes into the chloroform. LISON assumes that the color of the base is not necessarily involved in the color change brought about by addition of a chromotrope.

The failure of MICHAELIS to observe nucleic acid metachromasy was partly due, it is felt, to the pH-level at which all his experiments were performed—pH 4.6. WEISSMAN and his co-workers (1952) say that MICHAELIS did not work within the range of molar ratios of dye: nucleic acid phosphorus where nucleic acids are metachromatic, that is, between 0.4 and 1.4. In one of MICHAELIS' experiments, however, in an acetate buffer of pH 4.6 and ionic strength of 0.025, the dye: nucleic acid phosphorus ratios were between 0.001 and 10.8, including one solution having a ratio of 1.08. The solution that came closest to exhibiting a spectrum which could be called metachromatic had, in fact, a ratio of 0.108. On the other hand. the pH of 4.6 is below the range (pH 6–7) where WEISSMAN and his co-workers find nucleic acids to be metachromatic.

4. Temperature

Metachromasy is temperature-dependent, an increase in temperature suppressing the reaction (LISON 1935 a; BANK and BUNGENBERG DE JONG 1939: PÁLOS and KOCSÁN 1951; WEISSMAN, CARNES, RUBIN, and FISHER 1952). WEISSMAN and his co-workers were able to demonstrate metachromasy in a nucleic acid-toluidine blue solution by cooling the solution below 10^0 C. MICHAELIS (1950) displays the full spectral changes of metachromasy of agar-toluidine blue under the influence of temperature.

All temperature effects are completely reversible, both in solution and in tissue sections. As expected, if some component in a metachromatic system is altered by heat, the final level of metachromasy will differ from the original level (KELLY 1955 b). The sensitivity of metachromatic systems to thermal changes reflects the low bond-energies involved (MICHAELIS 1950; SYLVÉN 1954; WEISSMAN, CARNES, RUBIN and FISHER 1952).

5. Proteins

Proteins are the most important agents which might exert an influence on metachromasy, especially in any consideration of the state of chromotropes in tissues. Solution studies have shown that proteins, especially basic proteins, readily suppress metachromasy (BANK and BUNGENBERG DE JONG 1939; JAQUES 1943; HAMERMAN and SCHUBERT 1953, KELLY 1951, 1955 b). These solutions are discussed by KELLY in terms of their *potential* metachromasy, revealed by agents (heat, x-rays, proteolytic enzymes) which affect the protein or its state of combination with a chromotrope. HAMERMAN and SCHUBERT made the surprising observation that synovial fluid did not become metachromatic even when proteolytic enzymes were used to digest proteins to a level where they would not ordinarily suppress metachromasy. These authors concluded that synovial fluid hyaluronate may be in a native form incapable of reacting metachromatically with a dye.

FRENCH and BENDITT (1953) studied the influence of basic proteins on metachromatic staining of tissue sections, showing how such compounds can mask the metachromatic reaction and influence the pH optimum of dye-binding. The importance of *masking* as one explanation for variability in the staining of tissue sections has long been realized (HANSEN 1905).

E. Stoichiometry

The possibility of a stoichiometric relation for the metachromatic reaction has been questioned (MICHAELIS and GRANICK 1945). LISON (1935 a) says that the metachromatic reaction is not quantitative, that there is no stoichiometric relation between dye and chromotrope. For LISON, the intensity of "virage" depends on the reciprocal concentration of dye and chromotrope, not upon their relative concentrations.

What is known about the relation between dye-binding and metachromasy? While a number of investigators have studied dye-binding, it is felt that dye-binding is indeed a difficult thing to prove. Precipitates can readily form under conditions where the binding between dye and chromotrope is both loose and extremely heterogeneous (SYLVÉN 1954). Perhaps the most reliable studies of dye-binding in relation to metachromasy have been made by LEVINE and SCHUBERT (1952 b) and SCHUBERT and LEVINE (1953), who find a stoichiometric relation between dye-binding and concentration of chromotrope. This much has also been demonstrated by other authors (JAQUES, BRUCE-MITFORD, and RICKER 1947; MACINTOSH 1941; WALTON and RICKETTS 1954). In the investigations of LEVINE and SCHUBERT it was found that the concentration range where the major rise in dye-binding occurs is also the range where α- and β-peaks fall and the γ-peak rises. Above this point, however, important changes take place in the relative intensities at these wavelengths and such changes are independent of dye-binding. For example, the γ-band is suppressed by excess chromotrope without any effect on the binding of dye. The view is tentatively adopted by SCHUBERT and LEVINE that "the metachromatic color is due to binding of dye by chromotrope and is not due to dimerization or polymerization of the dye." There was no difference in binding that could serve to classify cationic dyes as metachromatic or non-metachromatic.

The metachromatic reaction is in some respects like the reactions of proteins with metal ions or dyes. There was at one time no "stoichiometry" of certain protein reactions before JAQUES LOEB and others showed that protein complexes must be defined within fairly definite environmental limits. For protein-metal ion interactions, stoichiometry means "... the number of ions which are bound by the macromolecule under a specified set of conditions" (KLOTZ 1952). In most metachromatic studies, the conditions are not adequately specified, an exception being the investigation of WEISSMAN, CARNES, RUBIN, and FISHER (1952) on nucleic acids. No particular case is made here for or against a stoichiometric relation in metachromasy. It is merely emphasized that such a relation should not be overlooked in future studies. The parameters of temperature, ionic strength and pH are obviously important, as well as purity of dye and of chromotrope. These, at least, must be specified.

Uses of Metachromasy

In one way or another, the metachromatic reaction has served as a tool in chemistry, medicine and histology. Only in histology and related fields has metachromasy been of sufficient importance to warrant more than cursory treatment in larger works. MICHAELIS' section on "Metachromasie" appeared in all three editions of EHRLICH's encyclopedia (MICHAELIS 1903, 1910, 1926). LISON (1936 a) summarized the knowledge on metachromasy as it was known shortly after he had demonstrated metachromasy *in vitro*. More recent information on metachromatic stains and metachromasy is to be found in a number of books on histochemical technique and histochemistry (CONN 1953; GATENBY and BEAMS 1950; GLICK 1949; GOMORI 1952; LILLIE 1954; LISON 1953; PEARSE 1953).

A. Histology and Cytology

A fairly sharp division exists in the application of metachromatic staining procedures to biological material. In *vital* or *supravital* staining, few of the principles learned from studies on solutions can be brought to bear on the living material. In the staining of *fixed* material, it is often possible to make greater use of the auxiliary techniques by which metachromasy is made more specific. Like all other histological studies, the direct and unqualified transference of information or techniques gained from *in vitro* studies to metachromatic staining is to be discouraged (LEVINE and SCHUBERT 1952 a; SINGER 1952; SYLVÉN and MALMGREN 1952).

1. Vital staining

General information on vital staining is found elsewhere in this Handbuch. Vital staining with metachromatic dyes is the subject of this section.

Quite naturally, the story of vital metachromatic staining is more concerned with cells than with tissues. PAFF and BLOOM (1949) observed metachromasy of the mast cells in cultures from dog tumors, including release of the cytoplasmic metachromatic substance via granules or vacuoles or by diffusion into the surrounding medium. COMPTON (1952) stained hamster mast cells supravitally. GROSSFELD (1954) observed metachromasy with toluidine blue in cultured fibroblasts and in supravitally stained epithelial cells of stomach, intestine and kidney. He points out that, while mast cells are metachromatic when vitally stained and also after fixation, fibroblasts and other cells are ordinarily only metachromatic when vitally stained. It is thought that the action of metachromatic dyes on non-living cells may be comparable to their action *in vitro* but that, in vital staining, additional factors associated only with the living cell are involved. LASFARGUES and DI FINE (1950) present a color plate of specific staining of fibroblasts, a red staining of the "Golgi zone" by a fraction (azure B?) extracted from an old sample of methylene blue. GROSSFELD also observed metachromasy to appear first in a region close to the nucleus.

The supravital staining of nerve fibers by methylene blue has been investigated by HALLER, STARR, and DAVENPORT (1949), aided by chromatography and spectrophotometry of dye solutions. While methylene blue was found satisfactory, toluidine blue and new methylene blue (better metachromatic dyes) did not stain any of the nervous elements. The specificity for nerve fibers is apparently associated with the $=N(CH_3)_2Cl$ group, some specificity being lost as one or both methyl groups are removed.

DUSTIN (1947) considers the vital staining of animal cells by metachromatic dyes to have several explanations. In some cases, no doubt, vital metachromasy has a histochemical significance (detection of sulfuric esters of high molecular weight). The mere concentration of dyes in vacuoles may lead to a metachromatic color, much as it would in ordinary solutions. In the case of certain vacuoles of erythrocytes and reticulocytes, DUSTIN feels that the difference between orthochromatic and metachromatic staining may represent only a difference in the physical state of ribonucleoprotein. However, he refers to two authors (JOKL, CESARIS-DEMEL) claiming to distinguish, or to have isolated, metachromatic "substances" from these cells.

The work of SPEK is highly important in any discussion of vital staining, being concerned with the relation between metachromatic staining and pH-determinations in living cells. References to Spek's extensive investigations are found in three of his papers (Spek 1940, 1942, 1944). He studied dye solutions, model systems of chromotropes and dyes, and vitally stained cells photometrically. He also injected chromotropes into cells which were previously or subsequently stained with a number of vital metachromatic dyes. Spek discusses fluorescence of metachromatic dyes. noting that Lison had failed to observe this phenomenon. The metachromatic color observed in concentrated dye solutions, according to Spek. is due to a molecular (non-dissociated) solution of the dye—especially is this so for dyes whose color bases are difficultly soluble in water. But basic dyes, says Spek, probably do not stain animal cells by a process of selective absorption of the dye base. In the case of dye-chromotrope reactions, the so-called "Fällungsmetachromasie" is an expression of the only special feature possessed by metachromatic dyes, namely, complex-formation. Spek does not accept Lison's original claim that metachromasy is a specific reaction for sulfuric esters, nor does he believe that a precipitation of chromotrope by metachromatic dye takes place under conditions of vital staining. Perhaps Spek's most important conclusion is that metachromasy and the colors of pH-indicators, in vital staining, have entirely different sources. Lison (1935 c) had earlier maintained that all basic vital stains capable of function as pH-indicators are also capable of metachromatic staining. He believed that use of such dyes as pH-indicators required proper choice of dye so that metachromatic color changes are not confused with those that might occur in a given pH range. The controversy between Spek and Lison is well summarized by Brachet (1947) in relation to other studies of vital staining. Some idea of the difficulties involved in making pH-determinations in living material are discussed by Heilbrunn (1952).

Metachromasy of acid dyes (Bordeaux red, Congo rubin) in relation to pH-indicators and vital staining is treated by Schulemann (1915), who also observed a "metachromasy" of colloidal gold when injected into mice and rabbits.

Czaja (1937) studied vital staining of plant cells, including metachromasy. and formulated a mechanism by which the living cell is stained. Briefly. the cell wall exerts an alkaline "membrane- or pore-effect" under which the dye-cation (or ionized dye base) is first adsorbed from the solution. The dye then passes into the cytoplasm or cell sap as the dye salt. Czaja's work will again be mentioned in the section on theories of metachromasy. There is other evidence that thiazine dyes may penetrate plant cells from highly ionized solutions (Brooks 1929), possibly in the form of the dye base (Irwin 1927).

Suspensions of nuclei are seen to be metachromatic with toluidine blue. under conditions common to both vital staining and the usual *in vitro* studies of metachromasy (Carnes, Weissman, and Rubin 1951). Similar results are obtained with smears or frozen sections examined in the hydrated state. Chromosomes in mitosis are metachromatic. The absorption maximum observed in this state is at 560 mμ, the γ-peak of toluidine blue in the presence of desoxyribonucleic acid.

Vital staining of the eggs of a number of lower animals revealed metachromasy to be rather widely distributed and often persistent under the conditions of permanent mounting (KELLY 1950, 1954).

2. Staining of sectioned material

In the preparation of more or less permanent histological preparations, various metachromatic staining procedures have enjoyed wide use for nearly eighty years. In the ordinary histological sense, where no claim is laid to *chemical* identification, the metachromatic reaction can be expected to yield consistent and reliable information on *structural* elements. The techniques involved are treated in the general histological or histochemical works cited above. Examples of the routine uses of metachromasy are found in the papers of BUNTING (1950), FRIBERG, GRAF, and ÅBERG (1951), MONTAGNA, EISEN, and GOLDMAN (1954), SYLVÉN (1941, 1945), and WISLOCKI, BUNTING, and DEMPSEY (1947 b). The paper of KRAMER and WINDRUM (1955) is recommended. The following outline of discussion is based in part on the technical divisions under which KRAMER and WINDRUM investigated metachromatic staining.

a) Fixation and pre-treatment

Only one special fixative has been strongly recommended for a tissue chromotrope and that is basic lead acetate for the mast cell granules (HOLMGREN and WILANDER 1937). This was modified by the addition of formaldehyde (SYLVÉN 1941), producing far superior fixation. Heavy metals are well-known as effective precipitants of acidic mucopolysaccharides but alone they are poor fixatives. If the lead-formaldehyde fixative is used, it should be prepared with freshly-boiled water and stored in a Pyrex bottle protected by soda lime, avoiding later exposure to air. Even so, if crystals appear in the section, PEARSE (1953) says that they may be removed by 0.1 N hydrochloric acid. It must be admitted that for the usual chromotropes, including mast cells, fixation is still largely a matter of opinion and that a number of common fixatives are entirely satisfactory (KRAMER and WINDRUM 1955; LILLIE 1949 b; MONTAGNA, EISEN and GOLDMAN 1954).

Solubility of chromotropes is an evident consideration. As in any other staining procedure, lipids demand aqueous media. BIENWALD (1939), FEYRTER (1936, 1942), FEYRTER and PISCHINGER (1942) and HERLANT (1943) used methods preventing the loss of weak metachromasy. MICHELS (1938) indicated the variable solubilities of mast cell granules. MONTAGNA, EISEN, and GOLDMAN (1954) observed no loss of metachromasy when human mast cells were kept in running water for 24 hours or longer, in contrast to the observation of JULÉN, SNELLMAN, and SYLVÉN (1950). PEARSE (1953) says that mucoproteins are quite soluble in water and even in 50–70% solutions of alcohol, acetone and dioxane. Absolute alcohol and acetone, while precipitating the mucoproteins, do not render them permanently insoluble.

b) Staining

Thionine used to be widely used for the staining of ordinary tissue chromotropes but it is less popular today than toluidine blue or azure A. Optically, there is less distinction between the orthochromatic and metachromatic states of thionine than between those of the other two dyes. Furthermore, Kramer and Windrum (1955) find that thionine contains a significant red component which indiscriminately stains many tissue components. The general purity of dyes should receive more attention in the future, following the observations of Kramer and Windrum and of Ball and Jackson (1953) and di Berardino (1954).

Pinacyanol has all the qualifications of a good metachromatic stain. although it has not enjoyed wide use (Bensley 1950; Proescher 1933; Sylvén 1954; Sylvén and Malmgren 1952). Proescher says that pinacyanol is also useful for amyloid metachromasy. The usual stains for amyloid are crystal violet or methyl violet of the triphenylmethane group (Highman 1946). Volutin may be stained metachromatically by toluidine blue in the ordinary way (Wiame 1947 a) as well as by a number of special methods which demonstrate the intense basophilia of volutin (Klein 1929; Lindegren 1947; Meyer 1904; Neumann 1932). Neumann lists a number of the older staining methods and reactions of volutin.

The *concentration* of dye is generally low, of the order of 0.1 % (Holmgren and Wilander 1937; Kramer and Windrum 1955). In addition to the low concentration of dye, Sylvén (1941, 1945) recommends that the solvent be 30 % alcohol and this is widely used. The use of low dye concentration and dilute alcohol is consistent with the recommendation of Lison (1935 a) that staining should be carried out under conditions where the orthochromatic color alone is stable. Hempelmann (1940) claims to distinguish epithelial and mesenchymal mucoproteins, based simply upon dilutions of metachromatic dyes. However, Lennox, Pearse, and Richards (1952) feel that the sharp distinction between such mucins is not as clear as it once was.

There is no firm recommendation on staining *time*, although it is typically 5–20 minutes for dye concentrations commonly used (Kramer and Windrum 1955; Sylvén 1941, 1945). For a 0.1 % toluidine blue solution in 30 % ethanol. rinsing in 95 % and absolute alcohol, and mounting in a synthetic resin. the staining of several different tissues was apparently at equilibrium within 30 minutes, since no further significant changes were noticed in the next 90 hours (personal observation). Thionine is maximally bound in the cytoplasm of nerve cells within 20 minutes (at pH 3.5) when binding is measured as the absorbance at 600 mμ (Koenig, Koenig, Eisenberg and Schildkraut 1954).

Buffers are often used in routine metachromatic staining (Bunting 1950: Windle, Rhines, and Rankin 1943) and the influence of pH has been investigated by Haynes (1928), Highman (1945), Landsmeer (1951), Montagna. Chase, and Melaragno (1951) and Montagna, Eisen, and Goldman (1954). Carnes and Forker (1954) find that the staining of amyloid by toluidine blue is abolished at pH 4.5, $\mu = 0.1$ while crystal violet staining of amyloid persists at pH 1.6, $\mu = 1.0$.

c) Post-treatment and mounting

In agreement with KRAMER and WINDRUM (1955), it is recognized that what happens to the section *after* staining may be more important than all previous treatment. The two routes that a stained section may follow, broadly speaking, involve either aqueous solvents or organic solvents.

It should be common practice to examine sections in the same solvent, pure water or otherwise, which is used for staining. While it is true that tissue metachromasy seen in an aqueous medium is often difficult to interpret, the observations made are instructive when compared with the pattern remaining after exposure to organic solvents and permanent mounting (cf. MICHAELIS 1947).

While LISON (1935 a) recommended glycerine-gelatine, levulose syrup, APÁTHY's gum syrup and canada balsam as mounting media for stained sections, it is doubtful that the special gums and syrups offer any advantage over resins for the normal metachromatic elements. KRAMER and WINDRUM (1955) use a synthetic resin (D. P. X.), balsam or dammar. Routinely, the writer used a synthetic resin whenever possible.

The special mounting media are still used for amyloid, whose metachromasy does not withstand, it is said, mounting in clarite or balsam (HIGHAM 1946). Various aqueous media or special procedures are used for weaker metachromatic elements, including lipids and mucus (BIENWALD 1939; FEYRTER 1936; FEYRTER and PISCHINGER 1942; HERLANT 1943; HESS and HOLLANDER 1947).

The metachromasy of nervous elements requires special treatment at various stages in the preparation of sections (CHANG 1938; NOBACK 1954; WISLOCKI and SINGER 1950; KRÜCKE 1939; WINDLE, RHINES, and RANKIN 1943).

One of the most controversial questions in the treatment of sections after staining concerns the influence of alcohol on metachromasy. LEHNER (1924) observed metachromasy in ground sections of bone, carefully dried and stained in a solution of thionine in absolute alcohol. He also observed metachromasy in mast cell granules in a preparation dried over several weeks in the unmounted condition, partly in sunlight. LEHNER felt that such observations raised doubt of Schmorl's idea that metachromasy required at least a trace of water. HANSEN (1908) believed that water was an absolute requirement for metachromasy. LISON (1935 a) heated a metachromatic preparation at 120^0 C. for one hour, destroying metachromasy and confirming HANSEN's belief. SYLVÉN (1954) says that metachromatic interaction requires the presence of water. PEARSE (1953) thinks that it is the presence of alcohol *and* water which is required for the destruction of metachromasy in tissue sections.

An experiment is cited by PEARSE in this connection. A section showing metachromasy due to an acid mucopolysaccharide is taken through 70% and absolute alcohol to xylene: the metachromasy is retained. If the latter section is now exposed to absolute alcohol, still the metachromasy remains. Instantaneous loss of metachromasy occurs, however, if one merely breathes on the section. This experiment was repeated by KRAMER and WINDUM (1955) with additional modifications. They agree with PEARSE in the observation that metachromasy persists after blotting and exposure to xylene and absolute alcohol, but feel that the blotting procedure is an inefficient method of dehydration. After drying stained sections of liver and cartilage over calcium chloride, the diffuse metachromasy in the liver disappeared but the cartilage metachromasy persisted. Metachromasy was *restored* to the liver section by breathing upon it. They conclude that alcohol acts by dehydration, that "false" metachromasy is destroyed by dehydration, and that "true" metachromasy survives under any mode of dehydration.

Water is extremely difficult to remove from acidic polysaccharides. Preparations of chondroitin sulfate, heparin, agar and gum arabic ordinarily contain 3-12% moisture as they are obtained commercially. Within a month or two in a calcium chloride desiccator, the moisture may be only 1-4%; it is not reduced below this level after several years in the desiccator. The desiccator is efficient in keeping heat-dried preparations practically moisture-free but there is a limit to the amount of moisture it will remove from any hygroscopic material.

In summary, the staining methods of SYLVÉN (1941, 1945) and KRAMER and WINDRUM (1955) are recommended for widest general use. These methods differ only in minor details. The only objection to either of the methods is in connection with "rinses" or relatively short exposures to one or more of the solvents. For staining procedures in general, not only metachromatic staining, it is not seen how any future development can be made unless empirical exposures are abandoned in favor of exposures assuring equilibrium between the tissue section and a given solution.

B. Histochemistry

A histochemical procedure is one in which a chemical compound or reaction is localized in terms of histological elements. Does the metachromatic reaction qualify as one of the *microscopic* histochemical methods? It has been said that "in a true histochemical test no differentiation must be employed" (PEARSE 1953). For metachromasy, this refers to the various means of dehydrating and mounting. KRAMER and WINDRUM (1955) take exception to PEARSE's statement, pointing out that it is standard chemical practice to insure the specificity of tests by properly selected steps, a practice much like "differentiation."

The histochemical use of metachromasy does not depend on the staining of tissue sections alone. General background studies include investigations on solutions (mentioned earlier in this paper), gels, smears or Coujard slides, and comparative studies, where metachromasy is correlated with other histochemical tests. Investigations of this type have been made by BRADEN (1955), COMPTON (1952), DAVIES (1952), GOMORI (1954), HALMI and DAVIES (1953), HASHIM (1952), LENNOX, PEARSE and RICHARDS (1952), MANCINI (1950), ROMANINI (1951), SCHLECHTER and CAMPANI (1948), SYLVÉN and MALMGREN (1952), WISLOCKI (1950) and WISLOCKI, BUNTING, and DEMPSEY (1947 b).

Under the conditions obtaining in tissue sections, it seems safe to assign metachromasy to *sulfated* compounds, probably sulfuric half-esters (GOMORI 1952; LISON 1935 a, 1936 a, 1953; PEARSE 1953; SYLVÉN and MALMGREN 1952). The so-called "beta" metachromasy (PEARSE 1953), which can be distinguished from the intense metachromasy of sulfated compounds, is probably due to nucleic acids (WISLOCKI, BUNTING, and DEMPSEY 1947 b). This weak metachromasy is not due to variations in technique since it appears in some nuclei and not others in the same section and it also co-exists in the same cell where the striking metachromasy characteristic of sulfated compounds is seen (KELLY 1954). While it is compelling to view the intense metachromasy of yeast cells as due to metaphosphate (WIAME 1949), there are claims of the presence of a sulfuric ester complex in the yeast cell (BELOZERSKY 1947). Thus, *phosphorylated* compounds usually exhibit weak metachromasy in animals, with the definite possibility, in some plants, of intense metachromasy equivalent to that of sulfated compounds.

Current evidence on the metachromasy of *carboxylated* compounds does not permit a definite statement. Some think that hyaluronic acid is responsible for metachromasy in sites (synovial fluid, vitreous body) where sulfated compounds are not found (WISLOCKI, BUNTING, and DEMPSEY 1947 b). It has been doubted that hyaluronate is ever in tissue concentrations high enough to produce visible metachromasy (MEYER 1951; SYLVÉN and MALMGREN 1952). HAMERMAN and SCHUBERT (1953) do not observe metachromasy in synovial fluid, not even after proteolytic digestion. On the other hand, these authors obtain metachromasy with sulfur-free hyaluronate at a concentration of 0.005%.

It is clear that the nature of the metachromatic reaction forbids sharp differentiation among chromotropes, that the degree of metachromasy merely indicates the relative charge density of a number of chromotropes (BANK and BUNGENBERG DE JONG 1939; SYLVÉN 1954). The specificity of the metachromatic reaction seems to depend, therefore, more upon the use of auxiliary techniques than upon the inherent selectivity of the metachromatic dyes for certain chromotropes. The use of dilute alcohol dehydrating solutions and mounting in special media are examples of empirical auxiliary techniques. Other attempts have been made to use more rational techniques for this purpose, including studies of those influences which interfere with metachromasy and lead to false assumptions concerning its specificity.

Selective extractions have been employed for the removal of basophilic components, especially ribonucleic acid. The extracting agents have included sodium chloride, perchloric acid, trichloroacetic acid, hydrochloric acid, nitric acid and sulfuric acid. References to the several methods are available in FISHER's (1952) paper. FISHER found that the mineral acids behaved like ribonuclease, without altering the results obtained with the Feulgen reaction, periodic acid-Schiff reaction, or metachromatic staining. ATKINSON (1952) believes that hydrochloric acid, perchloric acid and trichloroacetic acid are relatively non-specific and should not be used as substitutes for ribonuclease. These agents should particularly be avoided in the presence of extractable acid mucopolysaccharides.

Enzymes have played a dual role in relation to metachromasy. A number of enzymes have been used to remove the chromotrope itself and abolish metachromasy. Such enzymes are *hyaluronidase* (BUNTING 1950; CAMPANI and REGGIANINI 1950; COMPTON 1952; GRISHMAN 1952; MONTAGNA, CHASE, and MELARAGNO 1951; PENNEY and BALFOUR 1944; WISLOCKI, BUNTING, and DEMPSEY (1947 a, 1947 b), *chondromucinase* (LILLIE, EMMART, and LASKEY 1951), *diastase* (BENDITT and FRENCH 1953; LILLIE 1949 a), a *Neurospora enzyme* said to split 1, 4-linkages of galacturonic acid (COMPTON 1952), *ribonuclease* (WISLOCKI, BUNTING, and DEMPSEY 1947 b) and *desoxyribonuclease* (YASUZUMI, MORI. MATSUKURA, and MINAMINO 1950). There is considerable doubt on the specificities and purities of some of these enzymes, including the widely-used hyaluronidase (DURANS-REYNAL 1950, various papers; DAVIDSON and MEYER 1954; SYLVÉN and MALMGREN 1952). Enzymes attacking mucopolysaccharides are reviewed by FISHMAN (1951).

Enzymes have also been used to remove substances which suppress metachromasy. These "unmasking" enzymes are usually proteolytic enzymes, such as chymotrypsin, "collagenase," pepsin and trypsin (BENDITT and FRENCH 1953; FOLLIS 1951; GERSH and CATCHPOLE 1949; HAMERMAN and SCHUBERT 1953; KELLY 1951, 1955 b; YASUZUMI, MORI, MATSUKURA, and MINAMINO 1950). That one chromotrope of low degree of metachromasy may mask another more metachromatic element is shown by the use of ribonuclease to *increase* metachromasy in the cortical region of certain eggs (LANSING and ROSENTHAL 1949). A particularly interesting observation has been made on amyloid by KRAMER and WINDRUM (1955). Amyloid is not ordinarily metachromatic with toluidine blue. After peptic digestion. amyloid becomes metachromatic with this dye and the metachromasy is resistant to alcohol.

Various agents, usually *oxidizing agents*, have been employed to produce metachromasy (basophilia, at least) in histological elements which are not ordinarily metachromatic. These agents include chromic acid, peracetic acid, performic acid, periodic acid, permanganate, and bromine (DEMPSEY. SINGER, and WISLOCKI 1950; BURKL 1953; LILLIE, BANGLE, and FISHER 1954). LILLIE and his co-workers postulate, in the case of keratinous structures. that cystine is oxidized to cysteic acid and the sulfonic group of the latter is responsible for the observed metachromasy. Sulfuric acid is used by BIGNARDI (1940) to esterify mucoid substances in BRUNNER's glands. KRAMER and WINDRUM (1954) use sulfuric acid and other methods of sulfating tissue elements in order to convert them to chromotropes. The tissue elements are generally thought to be carbohydrates, though certain lipids (in the ground substance of the central nervous system, in the "kerasin" of GAUCHER's disease) may be sulfated if they survive the routine histological preparation.

Blocking agents are generally used in histochemistry to prevent a given reaction by specific interference (DANIELLI 1953, PEARSE 1953). FISHER and LILLIE (1954) are able to suppress the usual metachromasy of a number of elements by mild methylation with methanol in the presence of dilute hydrochloric acid. Subsequent demethylation restores metachromasy. It is particularly interesting that methylation rapidly removes metachromasy but only with difficulty suppresses the Feulgen and periodic acid-Schiff reactions. Certain "blocking agents" are especially important, since they may coexist with chromotropes in tissues and not only affect the character of metachromatic staining but also the physiological activities of the chromotropes. These agents are the proteins, especially basic proteins. HANSEN (1905) was aware of the masking effect in histological preparations. FRENCH and BENDITT (1953) find that a number of proteins, but especially histone, seriously alter the metachromatic staining of cartilage. Their conclusions are well worth repeating here:

(1) In the presence of basic protein the acidic character of the polysaccharide may be partly or completely masked.

(2) The pH dependence which is shown by mucopolysaccharides may be determined by associated protein and reflect the competition between the dye and the protein for acidic groups, rather than the dissociation characteristics of the acidic groups *per se*.

(3) Proteins present in crude tissue extracts applied to tissue sections may simulate hyaluronidase or chondroitinase activity by blocking the stainable groups.

The influence of basic proteins on metachromasy in solution is clearly shown by KELLY (1951, 1955 b).

Various *photometric procedures* show great promise in their application to the metachromatic reaction in cells and tissues. Microspectrophotometry in the visible region is discussed by LISON (1953), GLICK (1949), MOSES (1952) and POLLISTER and ORNSTEIN (1955). Special studies of metachromasy are not numerous. SPEK (1940) studied metachromasy in living cells by means of the photometer. The metachromasy of nucleic acids *in situ* has been extensively investigated by FLAX and HIMES (1952). It is pointed out by the latter authors that one difference between solutions and tissue sections is that absorption spectra for solutions are sums of orthochromatic and metachromatic spectra, while in tissue sections the spectrum is that of the bound dye alone. Under these conditions, both Beer's law and the Bouguer-Lambert law are obeyed. KELLY (1955 a) finds that the spectral characteristics of highly concentrated dye solutions are close to those of stained tissue sections. Whereas dichroism (p. 10) is ordinarily not a consideration of great importance in ordinary solutions, it must be reckoned with in the oriented systems investigated by means of the microspectrophotometer. It is not certain that the objections of COMMONER and LIPKIN (1949) with regard to spectrophotometry of histological material can yet be satisfactorily answered.

The "signature curves" characterizing various tissue components are obtained by colorimetric means (DEMPSEY, BUNTING, SINGER, and WISLOCKI 1947; DEMPSEY and SINGER 1946; SINGER 1952). By staining at various pH-levels, it is possible to show how dye-binding depends to some extent on the dissociation of cationic or anionic groups in tissue components. Thus, the metachromatic chromotropes (cartilage matrix, mast cells) continue to bind dye at pH-levels so low that other elements fail to stain. A number of factors interfere with simple interpretations of such curves (FRENCH and BENDITT 1953; SINGER 1954).

C. Chemistry

Numerous examples have been cited where metachromasy has been investigated for its own sake, in attempts to elucidate fundamental chemical principles. Metachromasy or metachromatic dyes have also been used as analytical tools.

The most common use of metachromatic dyes is in the assay of the anticoagulant, heparin. Only some of these methods are actually based on a metachromatic color (JAQUES, BRUCE-MITFORD, and RICKER 1947; JAQUES, MONKHOUSE, and STEWART 1949). Others depend, instead, on the extraction of the metachromatic complex by means of organic solvents and estimation of bound dye by colorimetric or photometric means (COPLEY and WHITNEY 1943; MACINTOSH 1941; SNELL and SNELL 1953; WALTON and RICKETTS 1954). THOMAS (1954) used a direct comparison method, where metachromatic activity was expressed as a ratio of amount of a standard heparin required to produce a given color to the amount of any chromotrope required to produce the same color. SYLVÉN (1951) uses metachromasy to follow the course of heparin extractions from mast cells. It is meaningless to state that anticoagulant and metachromatic activities of heparin are not related when the conditions of a "toluidine blue test" are not specified (WOLFROM, WEISBLAT, KARABINOS MCNEELY, and MCLEAN 1943), even though there may be evidence for such a statement (SNELLMAN, JENSEN, and SYLVÉN 1949).

The metachromasy of cholesterol and cholestanol sulfates is apparently more striking than that of heparin (GOLDENBERG and GOLDENBERG 1955). A test for steroid sulfates, including some that are not metachromatic, has been devised by GOLDENBERG (1955). It is an extraction method, using chloroform or ethylene dichloride, resembling the extraction methods for estimating heparin.

WIAME (1947 b) first reported the intense metachromasy of metaphosphate in solution and used a ratio to express the degree of metachromasy ($\mathfrak{E}_{530}/\mathfrak{E}_{630}$ for toluidine blue). This ratio seems generally applicable (HAMERMAN and SCHUBERT 1953; KELLY 1955 b). It might be stated as A_m/A_0, where A_m represents the absorbance at the wavelength of maximum metachromasy and A_0 represents the absorbance at the wavelength maximum of an extremely dilute solution. Other methods for the determination of metaphosphates are available (ALBAUM 1950; DAMLE and KRISHNAN 1954; SCHMIDT 1951).

Metachromatic dyes have been used to localize acid mucopolysaccharides on filter paper after chromatography (LEITNER and KIRBY 1954) or electrophoresis (RIENITS 1953).

Pinacyanol and rhodamine 6 G are used to determine the "critical concentration" of anionic detergents, where molecular dispersion gives way to the formation of molecular aggregates or micelles (CORRIN and HARKINS 1947). Similarly, anionic dyes like eosin, fluorescein, sky blue FF and acidified 2,6-dichloroindophenol are used for cationic detergents.

D. Medicine

Many uses of metachromasy in medicine are indirect, insofar as they concern stains which are used by the histologist or pathologist. Even in those cases where metachromatic dyes are used clinically, it is seldom "metachromasy" which is involved. Instead, advantage is taken clinically of the properties peculiar to metachromatic dyes, even though these properties are not fully elucidated.

Toluidine blue and other metachromatic dyes are effective in restoring the clotting time of blood in dogs exposed to x-rays, presumably by combining with a circulating anticoagulant (ALLEN, SANDERSON, MILLIAM, KIRSCHON, and JACOBSON 1948). Attempts to isolate heparin were only partly successful. In humans, toluidine blue produced slight nausea and vomiting in about 15% of patients (ALLEN, GROSSMAN, ELGHAMMER, MOULER, McKEEN, SANDERSON, EGNER, and CROSBIE 1949). GUTTERIDGE (1951) administered toluidine blue intravenously to 100 patients and observed transient nausea and occasional vomiting in 10% of cases. He remarks that ten times the recommended dosage of toluidine blue was administered by error to one patient in Boston: no ill effects were observed. Toluidine blue is apparently safely administered orally to control bleeding in oral surgery (TYLER 1954). It is difficult to see how toluidine blue acts in any other way than combination with a heparin-like substance in the blood. In this, the action of the dye closely resembles the action of protamine sulfate (CHARGAFF and OLSON 1938), a basic protein widely used in the control of heparin therapy. "Whether the bleeding factor in the blood is heparin or heparin-like, specific or nonspecific, is largely of academic interest, for the proper use of this method (protamine and toluidine blue) will save lives" (ODELL 1952).

Using metachromatic dyes to inhibit heparin *in vitro* in rabbit plasma, it was found that there is a critical level beyond which the dyes themselves exert an anticoagulant effect (HALEY and STOLARSKY 1950). This effect is unexplained other than as a postulated action on some system other than heparin.

An interesting experiment was performed by RILEY (1948). Noting that heparin enhanced the growth of tumor implants in mice and inhibited the growth of normal cells, RILEY wanted to know the effects of dyes calculated to bind heparin-like substances. For this purpose, he used the list of metachromatic dyes cited by LISON (1935 a). The most effective inhibitors of tumor growth were neutral red and neutral violet. Some effective dyes were not metachromatic and many were toxic. This is an important investigation, obscured somewhat by lack of knowledge on the purity and toxicity of dyes as they are used today.

Metachromatic dyes have been used to trace the distribution of administered dextran sulfates in tissues (MOWRY 1954). They have also been used to prevent the *in vitro* calcification of rachitic cartilage (MILLER, WALDMAN, and McLEAN 1952).

Theories of Metachromasy

A satisfactory theory of metachromasy must be consistent with general theories of staining (SINGER 1952, 1954) and with the known relations between color and chemical constitution (BRODE 1949; MELLON 1950). Such a theory must explain the metachromatic reaction in its several aspects —in solutions, gels and tissue sections—or else it must distinguish among them. There is no assurance that a single explanation will fit all these phenomena and, indeed, there may be evidence against such an assumption. A number of theories have been formulated to explain metachromasy specifically and advantage has been taken of these to help form the outline of this section. Other ideas and evidence, not dealing explicitly with the metachromatic reaction, are fitted into this framework.

EHRLICH (1877) distinguished the metachromatic color obtained with dahlia, primula, iodine violet, methyl violet and "purpurin" as a "Reaktionsfarbe," which is different from the mere increased intensity seen with safranin or fuchsin. Metachromasy was later called a "microchemical reaction," although no special explanation was offered (EHRLICH 1879 b). Following EHRLICH, several theories were advanced which have become of historical interest only, since subsequent information has clearly shown them to be inadequate. Certain hypotheses may correctly explain special phenomena but do not constitute general therories. Some few theories, while not currently supported, closely resemble or have given rise to modern ideas on the mechanism of the metachromatic reaction.

Until LISON (1935 a) had shown that metachromasy appears in a dye solution when a chromotrope is present, all theories were modulated by histological experience or investigations of dyes alone in solution. After LISON, the problem has become one of finding how closely tissue and solution studies may be compared and whether any general theory will fit the observations of both categories.

A. Optical Illusion

LEHNER (1924) and MICHELS (1938) refer to several authors who considered metachromasy to be a purely optical phenomenon, in the sense of chromatic aberration. LEHNER illustrates by simple experiments how comparable chromatic aberration may arise in the microscope and in dye solutions. It cannot be denied that such effects are to be expected when very small objects are examined and that the light-source, quality of optics and thickness of sections influence the colors observed. However, spectroscopic evidence long ago eliminated optical illusions as a fundamental explanation for metachromasy, whether occurring in solution or in tissue sections.

B. Impure Dyes

MICHAELIS (1903) and others (see LEHNER 1924) maintained that a chemically pure dye is able to color different tissue elements in different hues. We can only guess at the number of observations of "metachromasy" which actually were made by using impure dyes. HESCHL (1875), one of the discoverers of metachromasy, says that the dye he used was "ein Gemisch von Anilinblau mit Anilinroth." LISON (1935 a) lists sixteen early uses of metachromatic dyes, many of which are notorious mixtures and probably all impure to some extent. It seems permissible to eliminate speculation on the states of purity of dyes used by the older investigators. The metachromatic dyes used today are also impure but separation of their several fractions shows that some of the latter are still metachromatic (BALL and JACKSON 1953; KRAMER and WINDRUM 1955). Thus, while dye mixtures can simulate metachromasy, there is little doubt that a pure dye exhibits metachromasy.

C. Dye Bases

HANSEN (1908) and PAPPENHEIM (1906, 1910) were the chief proponents of the idea that metachromatic colors are explained by the formation of a dye base. It was observed that dye bases often differ in color from the corresponding salts but closely resemble the metachromatic colors. HANSEN thought that the color base, *preexisting* in a dye solution, was selectively bound in a loose chemical union with the proper substrate. The manner of binding is not clear; water is required and is involved in the sense of water-of-crystallization. PAPPENHEIM, claiming priority for the dye base theory, did not think that the dye base preexisted in the solution but was formed by interaction with the substrate. He was aware that this theory demands that the substrate have the properties of a strong base and he also knew that the usual substrate for a basic dye is acidic. PAPPENHEIM's explanation for the dissociation and subsequent binding of the dye base are not consistent with chemical theory. Furthermore, he believed it was the carbinol base which was involved. As LISON (1935 a) points out, when the carbinol base is obtainable at all, it is usually colorless.

MÖLLENDORFF (1924) showed that sometimes the dye base has the same color as a metachromatic reaction and sometimes the colors are different. Furthermore, there is a difference in resistance to alcohol for the metachromatic and dye base colors. LEHNER (1924) raises several objections to HANSEN's theory. HANSEN remarked on the resemblance of the colors of a concentrated solution to the metachromatic colors but LEHNER points out that dilution, not concentration, should favor the hydrolytic formation of the color base. Also, hydrolytic theories seem to be contradicted by observations of metachromasy in alcoholic solutions and in non-polar solvents. LEHNER, as well as LISON (1935 a), presumably extracted dye bases from aqueous solutions and found the solutions still metachromatic or capable of metachromatic staining. LISON and others have found that metachromasy occurs at pH-levels where the dye base does not exist. HAYNES

(1928) says that metachromatic staining by thiazine dyes is more pronounced between pH 5 and pH 7 than at higher or lower levels.

BANK and BUNGENBERG DE JONG (1939), while admitting that metachromasy resembles a shift in equilibrium favoring the dye base, observed that the color produced by the addition of strong base to toluidine blue is not reversible. A number of authors have stated that the formation of dye bases is not a fundamental part of the mechanism of metachromasy (HOLMES 1924, 1926 a; KELLEY and MILLER 1935 b; SPEK 1940). However, the dye base is indirectly involved by CZAJA (1930, 1934, 1935, 1937) in the vital meta-chromatic staining of plant cells. The cell wall behaves as an alkaline adsorbent of variable pore size which may form dye bases in the process of adsorbing the dye from solution. The dye salt is regenerated prior to passing into the cell sap or cytoplasm. CZAJA used ultrafilters to grade dyes according to particle size and related this to the passage of dye particles by the cell wall. The "ultrafiltration" by the cell wall governs the metachromatic colors assumed by plant cells. CERUTI (1940) thought that the α-, β-, and γ-bands of toluidine blue represented the dye cation, undissociated molecule, and undissociated dye base respectively.

It is unlikely that metachromasy, as it is ordinarily observed, is ex-plained by the formation of dye bases.

D. Tautomers

MICHAELIS (1903) first suggested that metachromasy depended on tautomer-ism, requiring the existence of at least one free amino group in the dye so that a labile hydrogen atom would be available for intramolecular translocation. Two forms of thionine were proposed:

(violet)   (red, metachromatic)

This was later viewed by MICHAELIS (1910) as an equilibrium between the undissociated metachromatic form and the ionic orthochromatic form. But the evidence of alcoholic solutions made even this seem unlikely, since the metachromatic color, which should appear when dissociation is suppressed, does not arise in alcohol. The alternative mechanism of polymerization was suggested at this time, based chiefly on freezing-point determinations. MICHAELIS also implicated the free base, saying that this form of the dye has a great tendency to polymerize. This is contrary to the observation of EPSTEIN, KARUSH, and RABINOWITCH (1941), who say that the free base (and trivalent cation) of thionine obeys Beer's law. It is curious that neither tautomerism nor polymerization is mentioned in the third edition of the Enzyklopädie (MICHAELIS 1926).

HOLMES (1924, 1926 a, 1926 b, 1927 b, 1928, 1929) made an exhaustive study of dyes, staining mechanisms, and metachromasy. Two later papers (HOLMES 1930; HOLMES, SCANLAN, and PETERSON 1932) are not directly concerned with staining but contain much information on dye spectra and spectrophotometry. HOLMES firmly believed in tautomerism as the best explanation of metachromasy. He rejected theories involving hydrolytic or electrolytic dissociation, aggregation, and effects of solvation. The tautomeric forms of MICHAELIS were thought to be chemically unsound and not sufficient to explain color changes, even if such tautomers existed. These tautomers, according to HOLMES, are impossible explanations for the metachromasy of triphenylmethane dyes (which was realized by MICHAELIS).

HOLMES says that metachromatic dyes exist in the solid state and in concentrated aqueous solutions as *addition products,* while in dilute aqueous and all alcoholic solutions they are *ammonium salts.* This is the basis for HOLMES' tautomerism. The relation of tautomerism to color effects depends on a change in nitrogen valency, $-N=$ to $-N\overset{\diagup}{\diagdown}$. There is no actual migration of atoms but a "redistribution of affinities" within the dye molecule. The relation of tautomerism to metachromatic staining depends on a differential absorption by the substrate. This selection is not explained beyond the statement that the tautomers must have distinctly different chemical and physical properties. Tautomers proposed for thionine were thought to be more consistent with chemical theory, although HOLMES admitted that they offered no better explanation of the *color* changes of metachromasy than the tautomers of MICHAELIS.

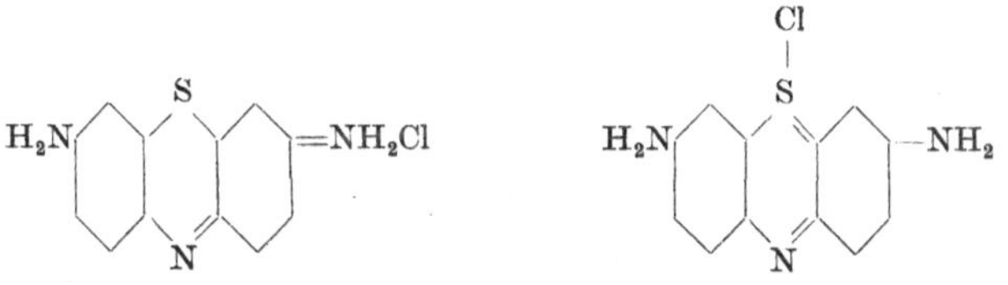

(paraquinoid)   (orthoquinoid)

Thionine

HOLMES studied brilliant cresyl blue in particular and offered a new type of tautomer, an addition product, which was also thought to be represented by the common zinc chloride double salt of methylene blue. These structures are shown below for brilliant cresyl blue, an oxazine dye, and for crystal violet, a triphenylmethane dye.

HOLMES (1924) displays a number of absorption spectra showing that, with concentration, there is a drop in absorbance at one wavelength (orthochromatic peak) and a simultaneous rise at some lower wavelength. LISON's (1935 a) curves show a progressive drop and a shift, which led him to say that the metachromatic shift could not be due to an equilibrium between tautomers. However, the curves shown by LISON are not typical of those obtained by other authors when a metachromatic dye is concentrated (BRODE 1955; RABINOWITCH and EPSTEIN 1941; MICHAELIS 1947), subjected to temperature changes (Brode 1955), or when a chromotrope is added to the dye (Kelly 1955 b; Michaelis 1947; Sylvén 1954).

The possibility of tautomerism is not excluded when considering only the equilibrium between α- and β-peaks, though this is ascribed to

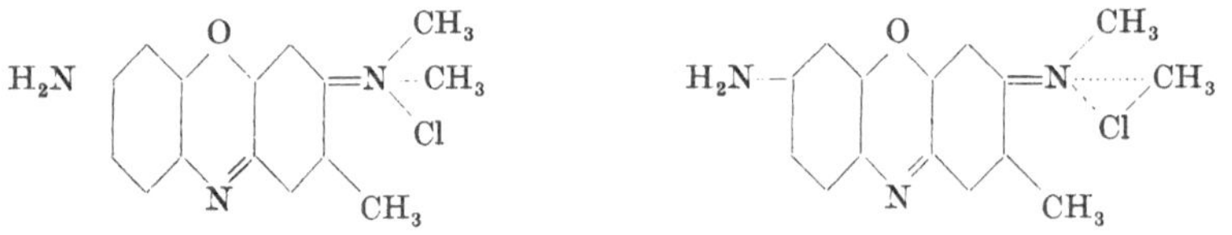

(blue)   (violet)

Brilliant cresyl blue

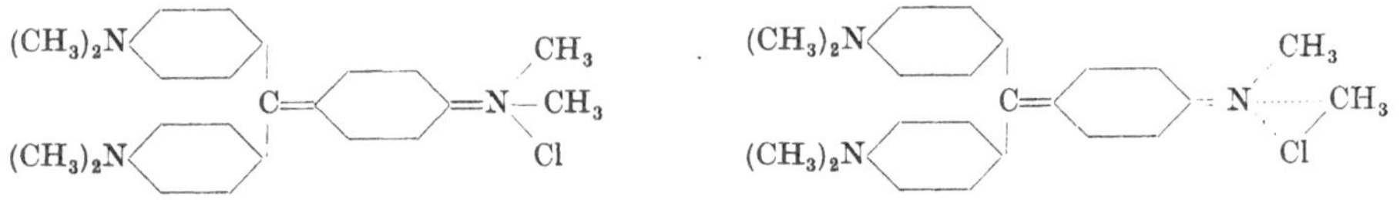

(blue)   (red)

Crystal violet

dimerization as well (Rabinowitch and Epstein 1941). The increased magnitude and complexity of spectral shifts associated with chromotrope metachromasy when γ-peaks arise seem to belie the occurrence of a simple tautomerism.

E. pH-Indicators

The possibility of confusion between metachromatic color changes and the changes of color indicators dependent upon pH has been mentioned elsewhere (p. 58). It is quite clear that metachromasy is broadly dependent on pH (Lison 1935 a), since the reaction shows a variable resistance to the addition of acid. It has been suggested by Clowes and Owen (1904) that methylene azure and the substance of the mast cell granules form a complex which is a pH-indicator.

F. Colloidal Theories

It is admitted that a number of theories are included under this heading for convenience only. Any idea which involves the formation of particles of colloidal dimensions is considered to be a colloidal "theory."

Schulemann (1915), referring to acid dyes only, thought that there were two modes of metachromasy. One was a coagulation of a colloidal solution (e. g., congo rubin) and the other was more like a crystallization (e. g., sulforhodamine). Schwarz and Hermann (1922) believed that metachromasy depended on the degree of dispersion of the absorbent and on its surface charge as conditioned by the ionic environment. The theory of Möllendorff (1924) included metachromasy as a special case of general staining. For Möllendorff, all staining processes involved either *saturation* (Durchtränkungsfärbung) or *precipitation* (Niederschlagsfärbung) colors. For many dyes, the tone of the precipitation color differs from that of the saturation color in the same way as the metachromatic color differs from the color of the same dye in alcoholic solution. Thus, metachromasy always represents the precipitation color. These precipitations typically take place at surfaces so that a granule, for example, might display the saturation or diffusion color in its interior at the same time that a precipitation color appeared at its surface.

Spek (1940) distinguishes color changes produced in pure dye solution from those produced by interaction of dye and chromotrope. In solutions, the color depends on the molecular state of the dye. A complex is formed between dye and chromotrope leading to what is called "Fällungsmetachromasie." The *critical concentration* of certain soap solutions where molecular dispersion gives way to aggregates or micelles is accompanied by color changes quite similar to metachromasy (Corrin and Harkins 1947).

Schlechter and Campani (1948) believe that metachromatic dyes must possess the unsaturated paraquinoidimine nucleus, permitting mesomerism between the ionic and undissociated forms of the dye. This is similar to the theories of Michaelis and Holmes. There is a decrease in the possibility of mesomerism when the dye is in the metachromatic (undissociated) form, a flocculent complex being formed with the chromotrope. Most metachromatic dyes do contain the paraquinoidimine structure. Thus, both thiazines and triphenylmethanes, whose metachromasy has always been difficult to reconcile under a single heading, fit the scheme of Schlechter and Campani. However, pinacyanol (a quinoline dye) and Bismarck brown (an azo derivative) are metachromatic and do not contain the paraquinoid structure.

G. Dimerization and Polymerization

While polymerization is perhaps only a special case of colloidal theories, it has received so much attention as an explanation of metachromasy that it is discussed separately. No other theory of metachromasy has been so widely accepted. It is, for example, the "textbook" theory of modern histology and histochemistry books.

The formation of dimers or polymers by metachromatic dyes was first suggested by Michaelis (1910). He had come to think that tautomerization was a less satisfactory explanation of metachromasy. For thionine, the polymerization represents an equilibrium,

$$2 \text{ (or 3) molecules } blue \rightleftharpoons 1 \text{ molecule } red$$

which is a mass law relation,

$$\frac{\text{conc. } blue}{\sqrt{\text{conc. } red}} = K.$$

For solutions of cationic dyes, deviations from Beer's law often mean association between cations and anions or between similar cations (polymerization) (Rabinowitch and Epstein 1941). In aqueous solution, the characteristic absorption changes with concentration may represent a reversible dimerization or the formation of higher polymers (SCHEIBE 1938). SCHEIBE, testing the application of mass law, obtained agreement only up to a certain concentration of dye and assumed that departure from the law at higher concentrations represented the formation of polymers. RABINOWITCH and EPSTEIN (1941), by making a different assumption on absorbance of the dye at infinite dilution, observed closer fit of their data with mass law. SHEPPARD and GEDDES (1944 b) criticize both theory and experimental procedure in relation to dimerization of dyes. They do not think that the mass law fails because the dimer hypothesis is incorrect. It is not the formation of new polymeric bands which leads to aberrations from the expected spectra. Instead, a vibrational excitation of the dimer occurs at higher concentrations and higher frequency satellites of the main electronic transition are thus enhanced. The possibility of "water of dimerization" is suggested, which is akin to HANSEN's (1908) idea that water is bound in a special way with the dye-chromotrope complex. SHEPPARD and GEDDES specifically reject the tautomer theories (SPEAS, MERRITT. HOLMES) in the case of "isothermal dilution change, for which only aggregation is likely."

MICHAELIS further developed the polymerization hypothesis (MICHAELIS 1944, 1947; MICHAELIS and GRANICK 1945). KOIZUMI and MATAGA (1953) adhere to the dimerization hypothesis but suggest that there may be two types of metachromasy: adsorption may influence the aggregation of the dye or it may influence the electronic state of the dye ion. Many cases of metachromasy, say these authors, are probably combinations of the two mechanisms.

LISON (1935 a) believed that the metachromatic reaction was a chemical phenomenon, not explained on a physico-chemical or optical basis and not dependent on polymerization or dispersity. His "théorie du displacement d'equilibre" expressed no more than the reversible equilibrium between the "normal" and "metachromatic" forms of the dye. However, LISON said that a satisfactory theory would combine (a) a colloidal change of the dye and (b) a simultaneous intramolecular change. Later in the same paper, this was more specifically restated as a simultaneous (a) polymerization and (b) tautomeric intramolecular transformation.

H. The Current Status

It may be assumed that polymerization is the current theory of metachromasy. What, now, is the evidence against such a hypothesis? It has already been mentioned that HOLMES (1924, 1926 a) was one of the first to assign a minor role to aggregation.

LISON and MUTSAARS (1950) think that there is only indirect evidence for dimerization or polymerization. WALTON and RICKETTS (1954) provide evidence that, in fact, neither polymerization of dye nor of substrate is involved in the production of metachromasy, which depends rather on the solubility properties of the dye-substrate complex. Their interpretation of experiments concerning the substrates is clear enough. The fact that only one dye molecule is bound by each anionic group in the substrate, however, does not interdict polymerization. The "polymerization" of MICHAELIS and others does not require a piling-up of dye ions at a single site. Optical interaction requires only that the dye ions be arranged in some sort of alignment, whether this be along the surfaces of large molecules, or in the presence of smaller molecules at a high concentration, or even when the dye alone is extremely concentrated. The polymerization may be indirect and mediated through the substrate, a situation satisfied by the one-to-one relation of dye and substrate-anion observed by WALTON and RICKETTS. An approximately "correct" molar ratio must also be secured for the existence of nucleic acid metachromasy (WEISSMAN, CARNES, RUBIN, and FISHER 1952).

There is evidence of a different kind showing that dimerization and polymerization may not be involved in metachromasy (LEVINE and SCHUBERT 1952 a, 1952 b; SCHUBERT and LEVINE 1953). These authors compare the simultaneous changes occurring in the several absorption maxima when a chromotrope is added to a metachromatic dye or when the dye alone is concentrated. The dimerization hypothesis cannot be supported. There is a qualitative agreement with the theory of polymerization but this is weak witness in the absence of other evidence. Furthermore, the meta-chromatic effect is to be distinguished from Beer's law deviation, since the α- and β-peaks shift in the same sense with the addition of a chromotrope but in the opposite sense when the dye is concentrated. The close relation between metachromatic ability of many dyes and their deviation from Beer's law is unexplained, unless this reflects the sensitivity of such dyes to external fields. The view is tentatively adopted by SCHUBERT and LEVINE that the metachromatic color is related to binding of the dye by the chromotrope, although there was no difference in binding that permitted the classification of dyes as metachromatic or non-metachromatic.

MICHAELIS (1950), in his last paper, questioned the validity of his own polymerization hypothesis. When an agar gel is stained with a very dilute solution of pinacyanol or toluidine blue, the metachromatic absorption bands still appear. MICHAELIS said that it is "unlikely that the dye should have been adsorbed in the form of polymolecular dye micelles, rather is a monomolecular distribution of the dye over the negatively charged sites the only reasonable assumption." He continued to think that nucleic acid behaves as if it depolymerizes the dye and that agar behaves as though it enhances polymerization. Dimerization and polymerization, says MICHAELIS, are just two factors possibly involved; the sensitivity of metachromatic dyes to external fields and the nature of substrates are other factors.

Various authors have suggested, especially in connection with theories of aggregation, that two steps are involved in the production of meta-chromasy. BANK and BUNGENBERG DE JONG (1939) said that electroadsorption is a primary factor and that aggregation is secondary, probably due to van der Waals forces. WEISSMAN, CARNES, RUBIN, and FISHER (1952) also believe that two types of bonds are involved. Polar bonds are first formed between the chromotrope anions and basic groups of the dye. These are sensitive to pH and ionic strength and their formation does not necessarily lead to metachromasy. Dye polymerization occurs through van der Waals forces between parallel rings of the dye molecules, producing the proper conditions for optical coupling which accounts for the spectral characteristics of metachromasy. These non-polar forces are weak, being affected by heat and alcohol. The interaction between dye and substrate and subsequent interaction between adjacent dye molecules are thus thought to be the two principal steps involved in metachromasy (SYLVÉN 1954).

The present status seems to be a general agreement on the primary binding of dye, which may or may not lead to metachromasy. If polymerization of the dye is *not* the secondary mechanism, is it reasonable to suppose that intramolecular changes, discussed earlier in this section, are responsible? Obviously, the dye-binding itself is important (SCHUBERT and LEVINE 1953). If it is true that the electronic transition states of dimers can be so altered by concentration of dye as to produce new spectral bands at lower wavelength (SHEPPARD and GEDDES 1944 a), then it may also be true that similar effects are exerted on single dye molecules. Some years ago, KELLEY and MILLER (1935 b) made the suggestion, based on an earlier color theory of MOIR, that a change in dye molecular volume might account for metachromatic color shifts. The volume change may depend on altered nitrogen bond angles. A chromotrope, in more firmly binding a dye, may cause the dye to absorb energy of higher frequency.

It cannot be suggested that any type of dye-substrate bond is unique with respect to metachromasy. However, it is conceivable that steric relations between dye and substrate can confer on the appropriate binding a diversity of character sufficient to explain the metachromatic reaction. Detailed knowledge of chromotrope structure, which is almost completely lacking, seems to be required for the further development of metachromatic theory. An adequate model of a chromotrope is available only in the case of desoxyribonucleic acid (WATSON and CRICK 1953).

Physiological Implications of Chromotropes

For practical purposes, a chromotrope is considered to be a polyacidic macromolecule. Certain properties peculiar to such compounds have been invoked to explain their physiological actions, and some of these properties may also account for the induction of metachromasy. The incidental color changes of a metachromatic reaction constitute direct visualization of a type of functional ability. With no desire to make more of this relationship than is reasonable, it is felt that the physiological activity of metachromatic substances should be illustrated by a limited number of examples. This section is devoted chiefly to sulfated or carboxylated compounds. The yeast chromotrope, volutin, requires only brief discussion in addition to that presented earlier. The nucleic acids are not treated in this short survey of chromotrope function, since they are so well known in other connections (CHARGAFF and DAVIDSON 1955, especially volume II). The unusual behavior of inorganic complexes with nucleic acids (NEUBERG 1949) is a striking illustration of the reactivity that may be shared by many chromotropes.

A. Some Properties of Polyacidic Colloids

Polymeric electrolytes are able to concentrate cations with some selectivity. KLOTZ (1952) refers to studies showing that substantial fractions of Na^+ and K^+ are complexed by polyphosphates. In a polyacrylate solution, as much as two-thirds of the sodium ions present may be associated with the polymer. The enormous magnitude of the electrostatic fields surrounding these polyacidic macromolecules may outweigh the influence of any special molecular configurations. KLOTZ says that "where these polyelectrolytes occur in living systems ... they must affect the distribution and physiological action of alkali-metal (as well as other) cations." When nucleic acid or, to a greater extent, heparin are present in salt solutions, the osmotic pressures of the solutions are considerably lower than is to be expected (JORPES 1946). It is suggested by JORPES that this immobilization of smaller ions is related to the mechanism by which some highly toxic compounds—heparin may contain as much as 45% sulfuric acid—are detoxified.

The effects on small ions may be compared to the binding of dye ions by chromotropes. Substrate configuration probably assumes greater importance in the case of dyes, conditioned by the greater size of the dye molecules themselves and their sensitivity to external fields. Especially when a chromotrope bears both heavily- and lightly-charged groups are the conditions favorable for metachromasy in tissue sections (BANK and BUNGENBERG DE JONG 1939). In a solution of chromotrope, there may exist micro-regions of high and low anionic density, the former favoring the binding of dye (LEVINE and SCHUBERT 1952 a).

The ability of mucopolysaccharides to form complexes with proteins is well known (EINBINDER and SCHUBERT 1951; JAQUES 1943; MEYER 1951). This binding of protein can lead to significantly altered properties, as JAQUES indicates, if the protein has a pronounced biological activity. The isoelectric point of proteins is shifted markedly by heparin (BEST and JAQUES 1949). Heparin inhibits a number of enzymes (JORPES 1946) and alters the enzymatic activity of lung thromboplastin by attaching to the protein carrier, replacing the phosphatide prosthetic group (CHARGAFF, ZIFF, and COHEN 1940). Similarly, heparin has been found to displace desoxyribonucleic acid from its histone complex in isolated liver cell nuclei (ROBERTS and ANDERSON 1951). COHEN (1942) tested a number of hydrophilic colloids for their ability to replace nucleic acid in tobacco mosaic virus. Unexpectedly, the virus was not only precipitated by several acidic colloids but was precipitated as paracrystalline needles. The hemocyanin of *Viviparus malleatus* was similarly crystallized. COHEN says that "these hydrophilic colloids may play a role in orientation of cytoplasmic particles and macromolecular solutes within the cell, in the sedimentation rate and clumping of blood cells and bacteria and in connection with other problems in which spatial organization of living tissues is important."

It has been suggested that metachromasy is merely a visual display of the ability of certain substrates to alter the energy states of bound compounds (BANK and BUNGENBERG DE JONG 1939). More broadly, these authors conceive of a colorless "metachromasy" when the bound compounds are not dyes. Perhaps the substrate and its inactive ligand form an active complex with entirely new properties of enzymatic or hormonal nature. Heparin, which is not itself an oxidizing agent, may transfer oxygen indirectly by binding and sensitizing dyes, as in a methylene blue-ascorbic acid system (PÁLOS and KOCSÁN 1951).

B. Integrity of Connective Tissue

The metachromatic reaction has been of service in the study of normal (ANGEVINE 1951; ASBOE-HANSEN 1951; BUNTING 1950; GERSH 1952; WISLOCKI, BUNTING, and DEMPSEY 1947 b) and pathological (ALTSCHULER and ANGEVINE 1949, 1951; BENNETT 1951; CAMPANI and REGGIANINI 1950; GERSH and CATCHPOLE 1949) connective tissue, especially the ground substance. Results obtained with metachromatic staining are to be compared with other histological or histochemical procedures (LILLIE 1952 a, 1952 b, 1952 c; see also p. 62). Interpretations of metachromatic reactions must be made in relation to knowledge of the chemistry and physiology of the connective tissues (BAKER and ABRAMS 1955; DURANS-REYNALS 1950; MEYER 1938, 1947, 1951, 1953; RAGAN 1953).

The mere distribution of mucopolysaccharides in the tissues and the simple mechanical properties of these compounds in sols or gels have suggested a number of functions. Thus, synovial fluid and mucous secretions seem to *lubricate* and *protect*, while the mesenchymal polysaccharides associated with ovarian granulosa cells may act as a *glue*, those of cartilage, blood vessels, nucleus pulposus and umbilical cord may help to maintain *elasticity* or *resiliency*, and the polysaccharides of the cornea and vitreous body may be related to the *transparency* of these structures (WISLOCKI, BUNTING, and DEMPSEY 1947 b). The transparency of a tissue, in turn, seems closely allied to the non-vascularity of that tissue (BUNTING 1950) and it was first suggested by MEYER (see RAGAN 1951) that the non-vascularization of the cornea is due to its mucopolysaccharide content, an explanation that may also apply to cartilage and amyloid.

Hyaluronic acid has a great capacity for hydration and may thus play a role in tissue water-binding (MEYER 1947). Sulfated mucopolysaccharides, on the other hand, may act as cation exchangers, a particularly attractive hypothesis in view of the anatomical locations of these compounds in basement membranes, dermis and walls of blood vessels (MEYER and RAPPORT 1951). The mucoitin sulfate of the gastric mucosa, says MEYER, may play a similar catalytic role in Na^+ shift or acid production of the stomach. NEUMAN, BOYD, and FELDMAN (1952) believe that the chondroitin sulfate of cartilage acts as a cation exchange resin. Interesting correlations can be drawn between metachromatic staining and calcification processes. The *in vitro* calcification of cartilage is prevented by metachromatic dyes (MILLER, WALDMAN, and McLEAN 1952). In bone, intense metachromasy is found in regions about to calcify (RUBIN and HOWARD 1950). While some (LEBLOND, BÉLANGER, and GREULICH 1955) think that chondroitin sulfate is actually involved in the binding of calcium in hard tissues, others (SOGNNAES 1955) assign to such a mucopolysaccharide the function of *prevention* of calcification. Certain experiments of SOBEL (1955) on cartilage, calcium, and toluidine blue suggest caution when relating metachromasy to states of calcification. In bone sections, calcium enhances metachromatic staining up to a certain concentration, then suppresses it. The same results are obtained in the staining of a synthetic chondroitin sulfate-collagen complex. However, calcium only interferes with metachromatic staining of chondroitin sulfate isolated from bone. SOBEL visualizes the process as a competition between calcium and the dye for the chondroitin sulfate complex. But there is also a structural effect of calcium on this complex, making the latter metachromatically more active. When the "new configuration" has been attained at a certain concentration of calcium, calcium then competes directly with the dye for the substrate and metachromasy is decreased.

Some of the properties of the connective tissues are thought to depend on the state of polymerization of mucoproteins or mucopolysaccharides of the ground substance (GERSH and CATCHPOLE 1949). Even if this were so, and the idea has been questioned (MEYER 1951), there is reason for thinking that metachromasy would not indicate the degree of polymerization (WALTON and RICKETTS 1954). LOEWI (1953) believes that depolymerization in cartilage favors calcification by increasing the number of sites available for calcium binding.

Evidence from studies on the reconstitution of collagen fibrils from solutions of collagen strongly suggests that mucoproteins are intimately related to the fibril formation (HIGHBERGER, GROSS, and SCHMITT 1951). In the electron microscope, it is seen that the axial spacing of the reconstituted collagen fibril depends on the mucoprotein concentration; a 640 Å period typical of native collagens or a 2000—3000 Å "LS fibril" ("long-spacing") period may be obtained. MORRIONE (1952) derived typical collagen fibrils from solutions when heparin was present in amounts of the order of 1 : 80,000.

Where does the ground substance originate? While it is generally assumed that the fibroblast forms the collagen fibril, the source of the ground substance is unknown. The fibroblast (GERSH 1952) and the mast cell (ASBOE-HANSEN 1954 b; LARSSON and SYLVÉN 1948) are proposed as manufacturers of the amorphous material. (Neither cell is found in cartilage.) SYLVÉN and his co-workers, with most investigators, believe that the sulfated components of connective tissue are derived from the mast cell. Attention is naturally drawn to the mast cell, unless there is some non-metachromatic precursor of metachromatic mucopolysaccharides in other cells, where metachromasy is rare (WISLOCKI, BUNTING, and DEMPSEY 1947 b). ASBOE-HANSEN, without denying that heparin is produced by the mast cell. thinks that this intensely metachromatic cell also produces hyaluronic acid. perhaps by way of a heparin-like precursor. This production is thought to be under hormonal regulation (ASBOE-HANSEN 1950, 1951). The important and interesting influence of several hormones (ASBOE-HANSEN 1954 a; RAGAN 1951) and vitamins (FELL 1953; PENNY and BALFOUR 1944) upon the connective tissue ground substance cannot be adequately discussed here. Among other observations, markedly altered metachromatic staining is often seen in the ground substance in association with hormonal or vitamin excess or deficiency. Except in unphysiological amounts (MEYER 1953), it is difficult to see how these agents would directly affect the ground substance.

C. Blood Clotting

Blood clotting (JAQUES 1954) and the anticoagulant, heparin (JORPES 1946), are only indirectly related to the subject of metachromatic staining. The most interesting common problem is the relation between anticoagulant activity and metachromatic activity. Many of the sulfated compounds listed in Table 5 have significant anticoagulant activity; the activity of carboxylated or phosphorylated compounds is typically of lower order.

Speaking only of anticoagulant activity, CHARGAFF, BANCROFT, and STANLEY-BROWN (1936) emphasized the importance of the sulfate group. No compound free of sulfur showed anticoagulant activity, except a phosphorylated cellulose of low activity. On the other hand, several sulfated compounds of high molecular weight were inactive. The requirements for anticoagulant power seemed to be (1) water solubility, (2) high molecular weight, and (3) the presence of sulfuric acid or other acid of similar strength. A direct relation between sulfur content and metachromatic activity of a number of natural and synthetic polymers has been clearly demonstrated (WALTON and RICKETTS 1954). In this study, WALTON and RICKETTS found that, in one series of sulfated glucose polymers (4000- 6000 glucose units), in vitro anticoagulant activity was shown by the entire series while in vivo activity was transitory when the number of glucose units was less than five, presumably because of rapid excretion in the urine. There is evidence that anticoagulant potency of heparin may be more closely related to its nitrogen content than to its sulfur content (JENSEN. SNELLMAN, and SYLVÉN 1948; WOLFROM and McNEELY 1945) and that neither sulfate nor "toluidine blue staining power" are true criteria of heparin activity (WOLFROM, WEISBLAT, KARABINOS, McNEELY, and McLEAN 1943). Heparin, when partially inactivated with respect to anti-clotting activity. still shows full metachromatic activity (JAQUES, BRUCE-MITFORD, and RICKER 1947). However, a table presented by JAQUES and his co-workers shows rough agreement between metachromatic and anticoagulant activities, considering the probable variance in purity of the compounds investigated.

Not long after the discovery of heparin, KING (1921) found some evidence for its existence in the uterine tissues of the pig. She thought the irregular yields of heparin from this source might have been due to masking by excess thromboplastic substances. There have been reports of the release of heparin-like substances into the bloodstream, yet proof of the presence of these substances in the blood has not been entirely convincing. In dogs irradiated with x-rays, the increased blood clotting time is restored by protamine sulfate and by several metachromatic dyes (ALLEN, SANDERSON, MILLIAM, KIRSCHON, and JACOBSON 1948). Attempts to isolate heparin from the blood of these dogs were only partly successful, the difficulty seeming to lie in the protein precipitation. In hemorrhage associated with a circulating anticoagulant, the failure to find free heparin in the blood has been explained on a different basis, namely, that the concentration of heparin is very critical with respect to the relative concentrations of heparin complement and thrombin (ZIFF and CHARGAFF 1940). The high heparin content of lung tissue and failure to obtain a metachromatic reaction with several fractions of this tissue, without special treatment, led CHARGAFF (1945) to say that "it is not impossible, therefore, that heparin was present in the protein complex in such a combination as to make it unavailable for the color test without previous drastic treatment."

Attention is directed to current interest in the ability of heparin to reduce lipemia. A *lipemia clearing factor* is apparently formed *in vivo* during peptone and anaphylactic shock (WORLEY and LEQUIRE 1955). After administration of protamine sulfate, almost no clearing effect is observed.

D. Cellular Activation and Inhibition

In spite of the numerous investigations on mucopolysaccharides, there is little known about the sources of these compounds, their metabolic roles, or their actions on cells. Tissue culture studies on bone (FELL 1953) and mast cells (PAFF and BLOOM 1949) represent those few studies directed at problems involving mucopolysaccharides at the cellular level. Another approach—that of the cellular physiologist who often works with *naturally* isolated cells—must not be overlooked.

An early appreciation of the role that heparin-like substances, or mucopolysaccharides, might play in cellular activities was shown by HEILBRUNN. For some years, HEILBRUNN (1952) has proposed that blood clotting and protoplasmic clotting are essentially alike and that clotting processes underlie certain fundamental aspects of cell performance. HEILBRUNN's book should be consulted for discussion of the implications of this theory and for references to the numerous investigations relating to it. At the present time, the attention of HEILBRUNN and his co-workers is directed at "substances that prevent the clotting of protoplasm and exert a powerful antimitotic action" (HEILBRUNN, CHAET, DUNN, and WILSON 1954). The ovaries of a variety of animals are good sources of these antimitotic agents, which are thought to be heparins or heparin-like substances. There is an interesting relation of these compounds to the activation of cells, where a type of "cellular homeostasis" is thought to operate (GOLDSTEIN 1953; HEILBRUNN 1952).

RUNNSTRÖM (1949, 1952) and his co-workers have also studied closely the processes of activation or fertilization of egg cells, particular attention being paid to the cell surface layers. The metachromasy in the cortical region or jelly coats of many egg cells represents sulfated polysaccharides which are capable of inhibiting blood clotting (IMMERS 1949; MONNÉ and HÁRDE 1951; MONNÉ and SLAUTTERBACK 1950; VASSEUR 1948 a, 1948 b). Considering the inhibitory effect of heparin-like substances on cell division (HEILBRUNN 1952) and their apparent association with normally arrested cells, it is interesting that basic proteins inhibit the jelly-coat substance or "fertilizin" of eggs (IMMERS 1949) and also incite cells to divide (WICKLUND 1947). However, a weak and temporary *inhibition* of cell division by basic proteins in a mouse carcinoma has also been reported (STEDMAN, STEDMAN, and PETTIGREW 1944).

The presence of metachromatic substances and mucopolysaccharides at the surface of many cells is to be contrasted with the absence of metachromasy in erythrocytes, although the erythrocyte surface may contain "mucoids" (MORGAN 1949). MORGAN suggests that the group receptors in the erythrocyte surface are mucoids, perhaps the "active patches" of DAVSON and DANIELLI, and that we should "recognize the existence of mucopolysaccharides as normal components of the surface structure of the red cell."

E. Metaphosphate (Volutin)

Volutin, the intensely metachromatic, intracellular component of yeast cells, bacteria and protozoa, is apparently a metaphosphate (ALBAUM 1950; DAMLE and KRISHNAN 1954; SMITH, WILKINSON, and DUGUID 1954; WIAME 1949). Its physiological role seems to be quite different from most metachromatic compounds of animals and higher plants, excepting nucleic acids. WIAME (1949) relates metaphosphate to the synthesis of nucleic acid. Metaphosphate may be an energy transmitter, at least, with no evidence that it is a direct phosphate donor (SCHMIDT, HECHT, and THANNHAUSER 1949). LINDEGREN (1947, 1948, 1951) has observed volutin appearing on the chromosomes and in the nucleolus of yeast cells before budding occurs and he also believes that volutin is involved in energy transfer.

In addition to metaphosphate, a mucopolysaccharide has been isolated from a capsulated strain of *Aerobacter aerogenes* (WARREN 1950). This polysaccharide contains no sulfur and is attacked by testicular hyaluronidase. The anticoagulant effect of hexametaphosphate is presumably due to the formation of a calcium complex (LARSON 1940). Metaphosphate is effective in precipitating (WIAME 1949) or crystallizing (PERLMAN 1938) proteins.

F. An Evaluation of the Metachromatic Reaction

Histologists and histochemists use the metachromatic reaction to localize substances in tissues, physiologists have used the reaction to detect or determine sulfated polysaccharides in body fluid, and chemists have studied metachromasy as a reaction significant in itself. All of these investigators have expressed direct opinions, at one time or another, concerning the value and meaning of metachromasy. In concluding this paper, it is felt that the current status of the metachromatic reaction may be summarized by a few statements recently made by investigators possessing considerable experience with various aspects of the method. These statements are to be considered against the background of evidence which has been presented throughout this review.

There are four basic methods for detecting acidic polysaccharide in tissues. The *Hale reaction*, which depends on the adsorption of ferric iron by more or less acidic compounds and subsequent development of the Prussian blue color, is virtually non-specific (BRADEN 1955; DAVIES 1952; LILLIE 1952 c). The *periodic acid-Schiff* (PAS) method, in which oxidation of 1, 2-glycol structures yields free aldehyde groups for coupling with the Schiff reagent, is chemically sound but its specificity for tissue components is questioned by an increasing number of authors and observations show a disturbing range of tissue components to be stained by the method. It might expected that sulfated compounds would not be PAS-positive (JORPES, WERNER, and ÅBERG 1948) but hyaluronic acid has also been found to react weakly or not at all (BRADEN 1955; DAVIES 1952; JEANLOZ 1950; MEYER and FELLING 1950). The *basophilia extinction* method for acidic compounds is a useful one and may be directly correlated with intensity of metachromatic staining (BRADEN 1955). There are exceptions to this correlation of basophilia and metachromasy (WISLOCKI, BUNTING, and DEMPSEY 1947 b), conflicting experience with the technique itself (DAVIES 1952), and definite possibility of serious interference by more or less basic tissue components (FRENCH and BENDITT 1953).

Metachromasy is distinctly more restricted in its typical histological occurrence than any of the three reactions mentioned above. Certain objections raised against basophilia extinction apply equally to metachromasy. Thus, "metachromasia and basophilia are of but limited value for distinguishing among the substances (acid mucopolysaccharides) in question," unless used in conjunction with more specific agents such as enzymes (BENDITT and FRENCH 1953). "Metachromasy, as in the case of basophilia, is therefore not an absolute matter..." (SINGER 1954). LISON (1953), now extending his definition of chromotropes to include acidic polysaccharides other than those which are sulfated, still maintains that the metachromatic reaction acquires histochemical significance only under conditions where the orthochromatic form of the dye is stable (see p. 49).

Among others, LEVINE and SCHUBERT (1952 a) have made valuable contributions to the theory of metachromasy by means of *in vitro* studies. They demonstrate the importance of such investigations but carefully point out that the results obtained are "not directly applicable to the interpretation of results of histological metachromatic staining," which is associated with "heterogeneous systems and is subject to far more complex influences." JAQUES (1954) has also studied *in vitro* metachromasy extensively, especially in relation to assay of anticoagulants, and he feels that "metachromasia is one of the most useful properties of heparin, since it can be used both chemically and histologically to identify heparin and related compounds."

SYLVÉN's (1954) views on the metachromatic reaction are clearly stated in the literature. These views are more recently summarized as follows: "Since metachromasia may be regarded as a reaction indicative of a certain 'free' (available) electro-negative surface charge density ... the old view about metachromasia as a specific histochemical reaction for the detection of certain chemical groups has to be abandoned. On the other hand, problems relating to the detection of various electro-negative substrates in cellular and tissue compartments become exceedingly difficult since these substrates usually form complexes with other electro-positive components, resulting in decrease of electro-negative charge density. The composition of such complexes may moreover vary during different conditions, and the quantitative ratios between negatively charged substrates and other molecules of the complexes will further influence the available surface charge densities. The only thing that seems certain is that a *positive* metachromatic reaction indicates that free negative charges are available and have reached a certain minimum density, *but a negative reaction does not exclude the presence of such metachromatic substrates.* This contention will place the metachromatic reaction in a group of physico-chemical reactions which have necessarily to be supplemented by other data before interpretations as to the chemical composition of the pertinent substrates are possible." [5] There is considerable support for SYLVÉN's statement (BENDITT and FRENCH 1953; FRENCH and BENDITT 1953; KELLY 1954, 1955 b; SINGER 1954; WALTON and RICKETTS 1954).

KRAMER and WINDRUM (1955) have performed a commendable service by reinvestigating some of the "more contentious aspects" of the metachromatic reaction, in order that a "valuable histochemical test in which there is no inherent defect, may be saved from falling into disrepute."

Acknowledgments

It is a pleasure to acknowledge the help of friends and colleagues who have contributed to this review, directly or indirectly, without in any way implicating them in responsiblity for errors or points of view. Dr. L. V. HEILBRUNN, Professor of Zoology in the University of Pennsylvania, offered many valuable suggestions for improvement of the manuscript. Doctors L. F. CAVAZOS and GORDON R. HENNIGAR, of the Departments of Anatomy and Pathology in the Medical College of Virginia, also read this paper carefully and I am deeply indebted to them. Many authors were kind enough to send their reprints and I take this opportunity of thanking them once again. For her characteristic accuracy and constant attention to the preparation of the manuscript, in the midst of innumerable other tasks, I wish particularly to thank Mrs. MABLE YOUNG, secretary in the Anatomy Department.

[5] Personal communication.

References

Asterisk (*) indicates author's initials not cited in original paper

ADLER, F. H., 1953: Physiology of the Eye, 2nd ed. New York.
ALBAUM, H. G., 1950: Phosphorylated compounds in *Euglena*. Arch. Biochem. **29**, 210—218.
ALLEN, J. G., M. SANDERSON, M. MILLIAM, A. KIRSCHON, and L. O. JACOBSON, 1948: Heparinemia (?). An anticoagulant in the blood of dogs with hemorrhagic tendency after total body exposure to roentgen rays. J. exp. Med. (Am.) **87**, 71—86.
— B. J. GROSSMAN, R. M. ELGHAMMER, P. V. MOULDER, C. L. McKEEN, M. SANDERSON, W. EGNER, and J. M. CROSBIE, 1949: Abnormal bleeding. II. Further clinical experience with toluidine blue and protamine sulfate. Surg., Gyn. a. Obst. **89**, 692—703.
ALLEN, R. D., 1951 a: Antimitotic substances secreted from eggs. Biol. Bull. (Am.) **101**, 214.
— 1951 b: The role of the nucleolus in spindle formation. Biol. Bull. (Am.) **101**, 214.
ALTSCHULER, C. H., and D. M. ANGEVINE, 1949: Histochemical studies on the pathogenesis of fibrinoid. Amer. J. Path. **25**, 1061—1077.
— — 1951: Acid mucopolysaccharides in degenerative disease of connective tissue, with special reference to serous inflammation. Amer. J. Path. **27**, 147—156.
ANDERSON, J. M., 1950: A cytological and cytochemical study of the male accessory reproductive glands in the Japanese beetle, *Popillia japonica* Newman. Biol. Bull. (Am.) **99**, 49—64.
ANGEVINE, D. M., 1951: Structure and function of normal connective tissue. Trans. First Conf. on Connective Tissues. J. Macy Jr. Found., 13—43.
ARGYRIS, T. S., 1954: The relationship between the hair growth cycle and the response of mouse skin to x-irradiation. Amer. J. Anat. **94**, 439—471.
ASBOE-HANSEN, G., 1950: A survey of the normal and pathological occurrence of mucinous substances and mast cells in the dermal connective tissue in man. Acta derm.-vener. (Schwd.) **30**, 338—347.
— 1951: On the mucinous substances of connective tissue. Thesis, Caroline Inst., Copenhagen (quoted in FRIBERG, GRAF, and ÅBERG, 1951).
— 1954 a: Hormonal effects on connective tissues. Trans. Fifth Conf. on Connective Tissues, J. Macy Jr. Found., 123—182.
— 1954 b: The mast cell. Internat. Rev. Cytol. **3**, 399—435.
ATKINSON, W. B., 1952: Differentiation of nucleic acids and acid mucopolysaccharides in histologic sections by selective extraction with acids. Science **116**, 303—305.
BABES, V., 1895: Beobachtungen über die metachromatischen Körperchen, Sporenbildung, Verzweigung, Kolben- und Kapselbildung pathogener Bakterien. Z. Hyg. usw. **20**, 412—437.
BAKER, B. L., and C. D. ABRAMS, 1955: The physiology of connective tissues. Annual Rev. Physiol. **17**, 61—78.
BALAZS, A., and H. HOLMGREN, 1950: The basic dye-uptake and the presence of a growth-inhibiting substance in the healing tissue of skin wounds. Exper. Cell Res. **1**, 206—216.
BALI, T., and J. FURTH, 1949: A transplantable splenic tumor rich in mast cells. Amer. J. Path. **25**, 605—625.
BALL, J., and D. S. JACKSON, 1953: Histological, chromatographic and spectrophotometric studies of toluidine blue. Stain Techn. **28**, 33—40.
BANK, O., und H. G. BUNGENBERG DE JONG, 1939: Untersuchungen über Metachromasie. Protoplasma **32**, 489—516.
BEAVEN, G. H., E. R. HOLIDAY, and E. A. JOHNSON, 1955: Optical properties of nucleic acids and their components. In CHARGAFF and DAVIDSON, Nucleic Acids, v. **1**, 493—553.
BELOZERSKII, A., 1945: The chemical nature of volutin. Microbiol. (USSR) **14**, 29—34 (Chem. Abstracts **40**, 616—617, 1947).
BELOZERSKY, A. N., 1947: On the nucleoproteins and polynucleotides of certain bacteria. Cold Spring Harbor Symp. Quant. Biol. **12**, 1—6.

BENDITT, E. P., and J. E. FRENCH, 1953: Histochemistry of connective tissue.
    I. The use of enzymes as specific histochemical reagents. J. Histochem. a.
    Cytochem. 1, 315—320.
BENNETT, G. A., 1951: Pathology of connective tissue, fibrinoid degeneration.
    Trans. First Conf. on Connective Tissues, J. Macy Jr. Found., 44—87.
BENSLEY, S. H., 1950: Histological studies of the reactions of cells and inter-
    cellular substances of loose connective tissue to the spreading factor of
    testicular extracts. Ann. N. Y. Acad. Sci. 52, 983—988.
BERDNIKOW, A., et C. CHAMPY, 1932: Recherches sur la substance mucoïde de la
    crête du coq. C. r. Soc. Biol. 106, 804—805.
BERGHE, L. VAN DEN, 1947: A cytochemical study of the "volutin granules" in
    Protozoa. J. Parasitol. 32, 465—466.
BERTALANFFY, F. D., and C. P. LEBLOND, 1953: The continuous renewal of the
    two types of alveolar cells in the lung of the rat. Anat. Rec. (Am.) 115,
    515—541.
BEST, C. H., and L. B. JAQUES, 1948: Heparin in blood clotting and thrombosis.
    Ann. N. Y. Acad. Sci. 49, 501—517.
BEVELANDER, G., 1952: Calcification in molluscs. III. Intake and deposition of
    Ca^{45} and P^{32} in relation to shell formation. Biol. Bull. (Am.) 102, 9—13.
— and P. BENZER, 1948: Calcification in marine molluscs. Biol. Bull. (Am.) 94,
    176—183.
BIENWALD, F., 1939: Zur Histochemie des Hirnanhanges. Virchows Arch. 303,
    577—587.
BIGNARDI, C., 1940: Cellule mucose e cellule mucoidi. IV. Esterificazione solfonica
    della sostanza mucoide e sua dimostrazione istochimica. Atti Soc. nat. e
    matem. (Modena) 71, 59—62.
— 1946: Considerazione critiche sulla reazione di metacromasi. Atti Soc. ital.
    Sci. Nat. 85, 160—170.
BRACHET, J., 1947: Embryologie chimique. Paris. (Eng. trans. by L. G. BARTH, 1950.)
BRADEN, A. W. H., 1955: The reactions of isolated mucopolysaccharides to several
    histochemical tests. Stain Techn. 30, 19—26.
BREWER, D. B., 1951: Myxoedema: an autopsy report with histochemical obser-
    vations on the nature of the mucoid infiltration. J. Path. a. Bacter. 63,
    503—512.
BRODE, W. R., 1949: Chemical Spectroscopy, 2nd ed. New York.
— 1955: Color and chemical constitution. Amer. Scientist 43, 259—284.
BROOKS, M. M., 1929: Penetration of thionine into Valonia. Amer. J. Physiol. 90,
    300.
BUNTING, H., 1950: The distribution of acid mucopolysaccharides in mammalian
    tissues as revealed by histochemical methods. Ann. N. Y. Acad. Sci. 52,
    977—982.
— G. B. WISLOCKI, and E. W. DEMPSEY, 1948: The chemical histology of human
    eccrine and apocrine sweat glands. Anat. Rec. (Am.) 100, 61—77.
BURKL. W., 1953: Zur Klassifizierung der Schleimdrüsensekrete in der Histologie.
    Z. Zellforsch. usw. 39, 74—84.
CAMPANI, M., and O. REGGIANINI, 1950: Observations in the experimental animal
    on the nature of the metachromatic ground substance in granulation tissue.
    J. Path. a. Bacter. 62, 563—568.
CARNES, W. H., and B. R. FORKER, 1954: The metachromasy of amyloid.
    J. Histochem. a. Cytochem. 2, 469—470.
— N. WEISSMAN, and P. S. RUBIN, 1951: Nuclear metachromasy. J. nat. Cancer
    Inst. 12, 240.
CASPERSSON, T., and K. BRANDT, 1941: Nucleotidumsatz und Wachstum bei Preß-
    hefe. Protoplasma 35, 507—526.
CERUTI, A., 1940: Ricerche sulle cause chimica e chimico-fisiche della meta-
    chromasia del bleu di toluidina e sue applicazioni all'interpretazione dei con-
    stituenti nucleari. Arch. botanico 16, 37—113.
CHANG, MIN-CHUEH, 1938: A formol-thionin method for the fixation and staining
    of nerve cells and fiber tracts. Anat. Rec. (Am.) 65, 437—441.
CHARGAFF, E., 1945: Cell structure and the problem of blood coagulation. J. biol.
    Chem. (Am.) 160, 351—359.
— F. W. BANCROFT, and M. STANLEY-BROWN, 1936: Studies on the chemistry of
    blood-coagulation. II. On the inhibition of blood clotting by substances of
    high molecular weight. J. biol. Chem. (Am.) 115, 155—161.

CHARGAFF, E., and J. N. DAVIDSON, 1955: The Nucleic Acids, 2 vols. New York.
— and K. B. OLSON, 1938: Studies on the chemistry of blood coagulation. VI. Studies on the action of heparin and other anticoagulants. J. biol. Chem. (Am.) **122**, 153—167.
— M. ZIFF, and S. S. COHEN, 1940: Studies on the chemistry of blood coagulation. X. The reaction between heparin and the thromboplastic factor. J. biol. Chem. (Am.) **136**, 257—264.
CLARK, W. M., 1948: Topics in Physical Chemistry. Baltimore.
CLOWES, G. H. A., and A. G. OWEN, 1904: The metachromatism of mast-cell granules and mucin. J. med. Res. (Am.) **12**, 407—431.
COHEN, S. S., 1942: The isolation and crystallization of plant viruses and other protein macromolecules by means of hydrophilic colloids. J. biol. Chem. (Am.) **144**, 353—362.
COMMONER, B., and D. LIPKIN, 1949: The application of the Beer-Lambert law to optically anisotropic systems. Science **110**, 41—43.
COMPTON, A. S., 1952: A cytochemical and cytological study of the connective tissue mast cell. Amer. J. Anat. **91**, 301—329.
CONN, H. J., 1953: Biological Stains. 6th ed. Geneva, N. Y.
CONSDEN, W., and W. BIRD, 1954: The carbohydrate of connective tissue. Nature **173**, 996—997.
COPLEY, A. L., and D. V. WHITNEY, 1941: The standardization and assay of heparin by the toluidine blue and azure A reactions. J. Labor. a. clin. Med. (Am.) **28**, 762—770.
CORNIL, V., 1875: Sur la dissociation du violet de méthylaniline et sa séparation en deux couleurs sous l'influence de certains tissues normaux et pathologiques, en particulier par les tissues en dégénérescence amyloïde. C. r. acad. Sci. **80**, 1288—1291.
CORRIN, M. L., and W. D. HARKINS, 1947: Determination of the critical concentration for micelle formation in solutions of colloidal electrolytes by the spectral change of a dye. J. Amer. Chem. Soc. **69**, 679—683.
COSTELLO, D. P., 1949: The relation of the plasma membrane, vitelline membrane, and jelly in the egg of *Nereis limbata*. J. gen. Physiol. (Am.) **32**, 351—366.
CURRAN, R. C., 1953: Observations on the formation of collagen in quartz lesions. J. Path. a. Bacter. **66**, 271—282.
CZAJA, A. Th., 1930: Untersuchungen über metachromatische Färbungen von Pflanzengeweben. I. Substantive Farbstoffe. Planta **11**, 582—626.
— 1934: Untersuchungen über metachromatische Färbungen von Pflanzengeweben. Planta **21**, 531—601.
— 1935: Der Membran- oder Poreneffekt des Absorptionsgewebes und seine physiologische Bedeutung. Planta **24**, 527—528.
— 1937: Untersuchungen über den Membraneffekt des Absorptionsgewebes und über die Farbstoffaufnahme in die lebende Zelle. Planta **26**, 90—119.
DAMLE, S. P., and P. S. KRISHNAN, 1954: Studies on the role of metaphosphate in molds. I. Quantitative studies on the metachromatic effect of metaphosphate. Arch. Biochem. a. Biophys. **49**, 58—70.
DANIELLI, J. F., 1953: Cytochemistry: A Critical Approach. New York.
DAVIDSON, E. A., and K. MEYER, 1954: Chondroitin, a new mucopolysaccharide. J. biol. Chem. (Am.) **211**, 605—611.
DAVIES, D. V., 1952: Specificity of staining methods for mucopolysaccharides of the hyaluronic acid type. Stain Techn. **27**, 65—70.
DEMPSEY, E. W., H. BUNTING, M. SINGER, and G. B. WISLOCKI, 1947: The dye-binding capacity and other chemo-histological properties of mammalian mucopolysaccharides. Anat. Rec. (Am.) **98**, 417—429.
— and M. SINGER, 1946: Observations on the chemical cytology of the thyroid gland at different functional stages. Endocrinology **38**, 270—295.
DEMPSEY, E. W., M. SINGER, and G. B. WISLOCKI, 1950: The increased basophilia of tissue proteins after oxidation with periodic acid. Stain Techn. **25**, 73—80.
DI BERARDINO, M., 1954: Dissimilar staining properties of purified and certified toluidine blue. Stain Techn. **29**, 253—256.
DOBELL, C., 1908: Notes on some parasitic protists. Quart. J. microsc. Sci. **52**, 121—138.

DUBOS, R. J., 1949: The nature and properties of the membranes of bacterial cells. Exper. Cell Res., suppl. 1, 192—203.

DURANS-REYNALS, F. (ed.), 1950: The ground substance of the mesenchyme and hyaluronidase. Ann. N. Y. Acad. Sci. 52, 943—1196.

DUSTIN, P., Jr., 1947: Ribonucleic acid and the vital staining of cytoplasmic vacuoles in animal cells. Symp. Soc. exper. Biol. (Cambridge) 1, 114—126.

EDLUND, T., and B. H. PERSSON, 1949: The influence of polar solvents on the metachromatic reaction *in vitro*. Experientia 5, 210—211.

EHRICH, W. E., 1953: Histamine in mast cells. Science 118, 603.

— J. SEIFTER, H. A. ALBURN, and A. J. BEGANY, 1949: Heparin and heparinocytes in Elephantiasis scroti. Proc. Soc. exper. Biol. a. Med. 70, 183—184.

EHRLICH, P., 1877: Beiträge zur Kenntnis der Anilinfärbungen und ihrer Verwendung in der mikroskopischen Technik. Arch. mikrosk. Anat. 13, 263—277.

*— 1879 a: Beiträge zur Kenntnis der granulirten Bindegewebzellen und der eosinophilen Leukocythen. Arch. Anat. u. Physiol. (Physiol. Abt.) 3, 166—169.

*— 1879 b: Über die specifischen Granulationen des Blutes. Arch. Anat. u. Physiol. (Physiol. Abt.) 3, 571—579.

EINBINDER, J., and M. SCHUBERT, 1951: Binding of mucopolysaccharides and dyes by collagen. J. biol. Chem. (Am.) 188, 335—341.

EPSTEIN, L. F., F. KARUSH, and E. RABINOWITCH, 1941: A spectrophotometric study of thionine. J. opt. Soc. Amer. 31, 77—84.

EWER, D. W., and J. HANSON, 1945: Some staining reactions of invertebrate mucoproteins. J. Roy. microsc. Soc. 65, 40—43.

FABER, M., 1949: The human aorta. Sulfate-containing polyuronides and the deposition of cholesterol. Arch. Path. (Am.) 48, 342—350.

FAWCETT, D. W., 1953: Experimental studies on the regeneration of mast cells. Anat. Rec. (Am.) 115, 305.

— 1954: Correlated cytological and pharmacological observations on the release of histamine by mast cells. Anat. Rec. (Am.) 118, 297.

FELL, H. B., 1953: The effect of vitamin A on organ cultures of skeletal and other tissues. Trans. Fourth Conf. on Connective Tissues, J. Macy Jr. Found., 142—184.

FERRY, J. D., 1939: Chemical and mechanical properties of two animal jellies. Biol. Bull. (Am.) 77, 331.

FEYRTER, F., 1936: Über ein sehr einfaches Verfahren der Markscheidenfärbung, zugleich eine neue Art der Färberei. Virchows Arch. 296, 645—654.

— 1942: Über chromotrope Lipide und Lipoproteide. Wien. klin. Wschr. 55, 461—463.

— und A. PISCHINGER, 1942: Über die Beziehungen zwischen den sogenannten chromotropen Lipoiden bzw. Lipoproteiden und den sogenannten Azetalphosphatiden in menschlichen Geweben. Wien. klin. Wschr. 55, 463—464.

FIESER, L. F., and M. FIESER, 1950: Organic Chemistry, 2nd ed., Boston.

FISHER, E. R., 1952: The destruction of cytoplasmic basophilia with mineral acids. Stain Techn. 28, 9—12.

— and R. D. LILLIE, 1954: The effect of methylation on basophilia. J. Histochem. a. Cytochem. 2, 81—87.

FISHMAN, W. H., 1951: Enzymatic hydrolysis of mucopolysaccharides. In SUMNER and MYRBÄCK, The Enzymes 1, part 2, 769—792.

FLAX, M. H., and M. H. HIMES, 1952: Microspectrophotometric analysis of metachromatic staining of nucleic acids. Physiol. Zool. 25, 297—311.

FOLLIS, R. H., Jr., 1951: Effect of proteolytic enzymes and fixation on metachromasia of skin collagen. Proc. Soc. exper. Biol. a. Med. (Am.) 76, 272—273.

FORTUNE, W. B., and M. G. MELLON, 1938: A spectrophotometric study of certain neutralization indicators. J. amer. Chem. Soc. 60, 2607—2610.

FRENCH, J. E., and E. P. BENDITT, 1953: The histochemistry of connective tissue: II. The effect of proteins on the selective staining of mucopolysaccharides by basic dyes. J. Histochem. a. Cytochem. 1, 321—325.

FRIBERG, U., W. GRAF, and B. ÅBERG, 1951: On the histochemistry of the mast cells. Acta path. et microbiol. scand. (Dän.) 29, 197—202.

FROMMHAGEN, L. H., M. J. FAHRENBACH, J. A. BROCKMAN Jr., and E. L. R. STOKSTAD. 1953: Heparinlike anticoagulants from Mollusca. Proc. Soc. exper. Biol. a. Med. (Am.) 82, 280—283.

GAGE, S. H., 1943: The Microscope, 17th ed. rev., Ithaca, N. Y.

GATENBY, J. B., and H. W. BEAMS, 1950: The Microtomist's Vade-Mecum (BOLLES-LEE), 11th ed. Philadelphia.

GERSH, I., 1952: Some functional considerations of ground substance of connective tissues. Trans. Second Conf. on Connective Tissues, J. Macy Jr. Found. 11—44.

— and H. R. CATCHPOLE, 1949: The organization of ground substance and basement membrane and its significance in tissue injury, disease and growth. Amer. J. Anat. 85, 457—522.

GIBSON, K. S., 1949: Spectrophotometry (200 to 1,000 millimicrons). U. S. Dept. Commerce, NBS Circ. 484, 1—48.

GLEGG, R. E., D. EIDINGER, and C. P. LEBLOND, 1954: Presence of carbohydrates distinct from acid mucopolysaccharides in connective tissue. Science 120, 839—840.

GLICK, D., 1949: Techniques of Histo- and Cytochemistry. New York.

GOLDENBERG, H., 1955: Use of basic dyes for the quantitative colorimetric estimation of steroid sulfate conjugates. Amer. Chem. Soc. Meetings, abstracts.

— and V. GOLDENBERG, 1955: Metachromasy of basic dyes induced by cholesterol and cholestanol sulfates. Amer. Chem. Society Meetings, abstracts.

GOLDSTEIN, L., 1953: A study of the mechanism of activation and nuclear breakdown in the *Chaetopterus* egg. Biol. Bull. (Am.) 105, 87—102.

GOMORI, G., 1952: Microscopic Histochemistry. Chicago.

— 1954: The histochemical behavior of acid mucopolysaccharides. J. Histochem. a. Cytochem. 2, 470.

GREEP, R. O. (ed.), 1954: Histology. New York.

— 1955: Recent advances in the study of the structure, composition, and growth of mineralized tissues. Ann. N. Y. Acad. Sci. 60, 541—806.

GRISHMAN, E., 1952: Histochemical analysis of mucopolysaccharides occurring in mucus-producing tumors. Cancer 5, 700—707.

GROSSFELD, H., 1954: Metachromasia in the living cell. Proc. Soc. exper. Biol. a. Med. (Am.) 86, 81—83.

GUILLIERMOND, A., G. MANGENOT, and L. PLANTEFOL, 1933: Traité de cytologie végétale. Paris.

GUTTERIDGE, N. M., 1951: The heparin protamine test and its clinical significance. Med. J. Australia 1, 403—408.

HALEY, T. J., and F. STOLARSKY, 1950: Effect of thiazine, oxazine and phenazine dyes on normal and heparinized rabbit plasma. Proc. Soc. exper. Biol. a. Med. (Am.) 73, 103—107.

HALLER, R. E., G. F. STARR, and H. A. DAVENPORT, 1949: Thiazin dyes in supravital staining of nerve fibers. Stain Techn. 24, 207—214.

HALMI, N. S., and J. DAVIES, 1953: Comparison of aldehyde fuchsin staining, metachromasia and periodic acid-Schiff reactivity of various tissues. J. Histochem. a. Cytochem. 1, 447—453.

HAMERMAN, D. J., and M. SCHUBERT, 1953: A quantitative study of metachromasy in synovial fluid and mucin. J. gen. Physiol. (Am.) 37, 291—300.

HANSEN, F. C. C., 1905: Untersuchungen über die Gruppe der Bindesubstanzen. 1. Der Hyalinknorpel. Anat. H. (Abt. I) 27, 535—820.

— 1908: Über die Ursachen der metachromatischen Färbung bei gewissen basischen Farbstoffen. Z. Mikrosk. 25, 145—153.

HARVEY, E. B., 1941: Vital staining of the centrifuged *Arbacia punctulata* egg. Biol. Bull. (Am.) 81, 114—118.

HASHIM, S., 1952: A survey of histochemical methods used for the identification of polysaccharides and their derivatives. Act anat. 16, 355—366.

HAYNES, R., 1927: Investigation of thiazine dyes as biological stains. I. The staining properties of thionin and its derivatives as compared with their chemical formulae. Stain Techn. 2, 8—16.

— 1928: Investigation of thiazine dyes as biological stains. II. Influence of buffered solutions on staining properties. Stain Techn. 3, 131—139.

HEILBRUNN, L. V., 1952: An Outline of General Physiology, 3rd ed. Philadelphia.

— A. B. CHAET, A. DUNN, and W. L. WILSON, 1954: Antimitotic substances from ovaries. Biol. Bull. (Am.) 106, 158—168.

— and W. L. WILSON, 1950: Effect of bacterial polysaccharide on cell division. Science 112, 56—57.

HEMPELMANN, L. H., 1940: Staining reactions of mucoproteins. Anat. Rec. (Am.) 78, 197—206.

HENRICI. A. T., 1930: Molds, Yeasts, and Actinomycetes. New York.

HERLANT, M., 1943: Recherches sur la localisation histologiques des hormones gonadotropes femelles au niveau de l'hypophyse antérieure. Arch. Biol. (Fr.) **54**, 225—357.

HERWERDEN, M. A. VAN, 1917: Over den aard de beteekenis der volutine in gistcellen. Versl. d. Afdeeling Natuurk. (Akad Konink. Amsterdam) **25**. 1445—1463.

*HESCHL, 1875: Eine hübsche a-vista-Reaktion auf Amyloid degenerirte Gewebe. Wien. med. Wschr., **32** (Jahrg. 25), 714—715.

HESS, M., and F. HOLLANDER, 1947: Permanent metachromatic staining of mucus in tissue sections and smears. J. Labor. a. clin. Med. (Am.) **32**, 905.

HIGHBERGER, J. H., J. GROSS, and F. O. SCHMITT, 1951: The interaction of mucoprotein with soluble collagen; an electron microscope study. Proc. Nat. Acad. Sci. **37**, 286—291.

HIGHMAN, B., 1943: Staining of mucus with buffered solutions of toluidine blue O. thionin and new methylene blue N. Stain Techn. **20**, 85—87.

— 1946: Improved methods for demonstrating amyloid in paraffin sections. Arch. Path. (Am.) **41**, 559—562.

HIMES, M. H., and M. H. FLAX, 1950: Staining of nucleic acids by azure-phthalate. Anat. Rec. (Am.) **108**, 539—540.

HIRAI, E., 1949: On test-cell of egg of the ascidian, *Cynthia roretzi* Drasche and its secretion. Zool. Mag. **58**, 205 (in Japanese, quoted from KUSA, 1954).

HOLMES, W. C., 1924: The influence of variation in concentration on the absorption spectra of dye solutions. Indust. a. Eng. Chem. **16**, 35—40.

— 1926 a: The chemical nature of metachromasy. Stain Techn. **1**, 116—122.

— 1926 b: The mechanism of stain action with basic dyes. Stain Techn. **1**, 98—102.

— 1927 a: Subsidiary dyes in methylene blue. Stain Techn. **2**, 71—73.

— 1927 b: Tautomerism of aminated dyes with variation in concentration. Amer. Dyestuff Reporter **16**, 429—432.

— 1928: The tautomerism of brilliant cresyl blue. J. amer. chem. Soc. **50**. 1989—1993.

— 1929: The mechanism of staining: The case for the physical theories. Stain Techn. **4**, 75—80.

— 1930: The absorption spectra of dyes. Internat. Crit. Tables **7**, 173—211.

— J. T. SCANLAN, and A. R. PETERSON, 1932: The visual spectrophotometry of dyes. U.S. Dept. Agric., Techn. Bull. **310**, 1—41.

HOLMGREN, H., 1949: Metachromasia in growing tissue. Exper. Cell Res. suppl. **1**. 378—379.

— and O. WILANDER, 1937: Beitrag zur Kenntnis der Chemie und Funktion der Ehrlichschen Mastzellen. Z. mikrosk.-anat. Forsch. **42**, 242—278.

— and G. WOHLFART, 1948: Mast cells in experimental rat sarcomas. Cancer Res. **7**, 686—691.

HOLTFRETER, J., 1946: Experiments on the formed inclusions of the amphibian egg. II. Formative effects of hydration and dehydration on lipid bodies. J. exper. Zool. **102**, 51—108.

IMMERS, J., 1949: On the heparin-like effect of fertilizin from sea-urchin eggs on blood coagulation and the modifying action of basic proteins and cephalin on the effect. Arkiv Zool. **42 A**, No. 6, 1—9.

IRWIN, M. 1927: On the nature of the dye penetrating the vacuole of *Valonia* from solutions of methylene blue. J. gen. Physiol. (Am.) **10**, 927—947.

JAKUS, M. A., 1945: The structure and properties of the trichocysts of *Paramecium*. J. exper. Zool. **100**, 457—485.

JAQUES, L. B., 1943: The reaction of heparin with proteins and complex bases. Biochem. J. (Brit.) **37**, 189—195.

— 1954: Blood clotting and hemostasis. Annual Rev. Physiol. **16**, 175—214.

— M. BRUCE-MITFORD, and A. G. RICKER, 1947: The metachromatic activity of heparin. Rev. Canadienne Biol. **6**, 740—754.

— F. C. MONKHOUSE, and M. STEWART, 1949: A method for the determination of heparin in blood. J. Physiol. (Brit.) **109**, 41—48.

— and E. KEERI-SZANTO, 1952: Heparinase. II. Distribution of enzyme in various tissues and its action on natural heparins and certain synthetic anticoagulants. Canad. J. med. Sci. **30**, 353—359.

JEANLOZ. R., 1950: Hotchkiss reaction and structure of polysaccharides. Science **111**, 289.

Jensen, R., O. Snellman, and B. Sylvén, 1948: On the inhomogeneity of commercial heparin from the physicochemical point of view. J. biol. Chem. (Am.) 174, 265—271.

Jorpes, J. E., 1946: Heparin in the Treatment of Thrombosis. 2nd ed. London.

— B. Werner, and B. Åberg, 1948: The fuchsin-sulfurous acid test after periodate oxidation of heparin and allied polysaccharides. J. biol. Chem. (Åm.) 176, 277—282.

Julén, C., O. Snellman, and B. Sylvén, 1950: Cytological and fractionation studies on the cytoplasmic constituents of tissue mast cells. Acta physiol. scand. 19, 289—305.

Jürgens, R., 1875: Eine neue Reaction auf Amyloidkörper. Virchows Arch. 65, 189—196.

Kelley, E. G., and E. G. Miller, Jr., 1935 a: Reactions of dyes with cell substances. I. Staining of isolated nuclear substances. J. biol. Chem. (Am.) 110, 113—118.

— — 1935 b: Reactions of dyes with cell substances. II. The differential staining of nucleoprotein and mucin by thionine and similar dyes. J. biol. Chem. (Am.) 110, 119—140.

Kelly, J. W., 1950: The localization of a metachromatic substance in the Chaetopterus egg. Protoplasma 39, 386—388.

— 1951: Effects of x-ray and trypsin on the metachromasy of heparin-basic protein combinations. Biol. Bull. (Am.) 101, 223.

— 1954: Metachromasy in the eggs of fifteen lower animals. Protoplasma 43, 329—346.

— 1955 a: Spectrophotometry of the metachromatic reaction in solution and in tissue sections. Anat. Rec. (Am.) 121, 319.

— 1955 b: Suppression of metachromasy by basic proteins. Arch. Biochem. a. Biophys. 55, 130—137.

Kelsall, M. A., and E. D. Crabb, 1954: Effects of x-irradiation on mast cells and mucous secretion in Brunner's glands. Anat. Rec. (Am.) 118, 393.

King, J. L., 1921: A study of the anticoagulating substance in the mucous membrane of the uterus. Amer. J. Physiol. 57, 444—453.

Klein, G., 1929: Praktikum der Histochemie. Wien, Berlin.

Klemperer, P., 1950: The concept of collagen diseases. Amer. J. Path. 26, 505—519.

Klotz, I. M., 1952: Some metal complexes with proteins and other large molecules. In Barron, Modern Trends in Physiology and Biochemistry, 427—451.

Koenig, H., R. Koenig, J. Eisenberg, and D. Schildkraut, 1954: Studies of basic dye-nucleic acid interaction. Anat. Rec. (Am.) 118, 320.

— — — — 1954: Studies of basic dye-nucleic acid interaction in fixed nervous tissue. J. Histochem. a. Cytochem. 2, 448.

Koizumi, M., and N. Mataga, 1953: Metachromasy of rhodamine 6 G produced by polyvinyl sulfate. J. amer. chem. Soc. 75, 483—484.

Koksal, M., 1953: Extraction of a heparin-like substance from mast cell granules in mouse connective tissue. Nature 172, 733—734.

Kramer, H., and G. M. Windrum, 1953: Metachromasia after treating tissue sections with sulphuric acid. J. clin. Path. 6, 239—240.

— — 1954: Sulphation techniques in histochemistry with special reference to metachromasia. J. Histochem. a. Cytochem. 2, 196—208.

— — 1955: The metachromatic staining reaction. J. Histochem. a. Cytochem. 3, 227—237.

Krücke, W., 1939: Die mucoide Degeneration der peripheren Nerven. Virchows Arch. 304, 442—463.

Kurnick, N. B., 1950: Methyl green-pyronin. I. Basis of selective staining of nucleic acids. J. gen. Physiol. (Am.) 33, 243—264.

— 1952: Histological staining with methyl-green-pyronin. Stain Techn. 27, 233—242.

Kusa, M., 1954: The cortical alveoli of salmon egg. Annot. Zool. Jap. 27, 1—6.

Landsmeer, J. M. F., 1951: Some colloid chemical aspects of metachromasia. Influence of pH and salts on metachromatic phenomena evoked by toluidine blue in animal tissue. Acta physiol. et pharmacol. Neerl. 2, 112—128.

Lansing, A. I., and T. B. Rosenthal, 1949: Ribonucleic acid at cell surfaces and its possible significance. Biol. Bull. (Am.) 97, 263.

Larson, C. E., 1940: Use of sodium hexametaphosphate as an anticoagulant. Proc. Soc. exper. Biol. a. Med. (Am.) 44, 554—555.

LARSSON, G., and B. SYLVÉN, 1948: The mast-cell reaction of mouse skin to some organic chemicals. Cancer Res. 7, 680—685.

LASFARGUES, E., and J. DI FINE, 1950: Specific vital staining of the Golgi zone in tissue culture with Azure B. Anat. Rec. (Am.) 106, 29—33.

*LAVERAN, A., and T. MESNIL, 1901: Sur la nature centrosomique du corpuscle chromatique postérieur des trypanosomes. C. r. Soc. Biol. 53, 329—331.

LEBLOND, C. P., L. F. BÉLANGER, and R. C. GREULICH, 1955: Formation of bones and teeth as visualized by radioautography. Ann. N.Y. Acad. Sci. 60, 629—659.

LEHNER, J., 1924: Das Mastzellen-Problem und die Metachromasie-Frage. Erg. Anat. 25, 67—184.

LEITNER, J. G., and G. P. KERBY, 1954: Staining of acid mucopolysaccharides after chromatography on filter paper. Stain Techn. 29, 257—259.

LENNOX, B., A. G. E. PEARSE, and H. G. H. RICHARDS, 1952: Mucin-secreting tumours of the skin: with special reference to the so-called mixed-salivary tumour of the skin and its relation to hidradenoma. J. Path. a. Bacter. 64, 865—880.

LEVINE, ANNE, and M. SCHUBERT, 1952 a: Metachromasy of thiazine dyes produced by chondroitin sulfate. J. amer. chem. Soc. 74, 91—97.

— — 1952 b: Metachromatic effects of anionic polysaccharides and detergents. J. amer. chem. Soc. 74, 5702—5706.

LILLIE, R. D., 1949 a: On the destruction of cytoplasmic basophilia (ribonucleic acid) and of the metachromatic basophilia of cartilage by the glycogen splitting enzyme malt diastase: a histochemical study. Anat. Rec. (Am.) 103, 611—633.

— 1949 b: Studies on the histochemistry of normal and pathological mucins in man and in laboratory animals. Bull. Internat. Assoc. Med. Museums 29, 1—55.

— 1952 a: Connective tissue staining. Trans. Second Conf. on Connective Tissues, J. Macy Jr. Found. 11—37.

— 1952 b: Histochemistry of connective tissues: collagen, reticulum, basement membranes, sarcolemma, ocular membranes. Lab. Invest. 1, 30—45.

— 1952 c: Staining of connective tissue. Amer. med. Assoc. Arch. Path. 54, 220—233.

— 1954: Histopathologic Technic and Practical Histochemistry. New York.

— R. BANGLE, and E. R. FISHER, 1954: Metachromatic basophilia of keratin after oxidation-cleavage of disulfide bonds. J. Histochem. a. Cytochem. 2, 95—102.

— E. W. EMMART, and A. M. LASKEY, 1951: Chondromucinase from bovine testis and the chondromucin of the umbilical cord. Amer. med. Assoc. Arch. Path. 52, 363—368.

LINDBERG, O., and L. ERNSTER, 1954: Chemistry and physiology of mitochondria and microsomes. In HEILBRUNN and WEBER, Protoplasmatologia 3, 1—136.

LINDEGREN, C., 1945: An analysis of the mechanism of budding in yeasts and some observations on the structure of the yeast cell. Mycologia 37, 767—780.

— 1947: The function of volutin (metaphosphate) in mitosis. Nature 159, 63—64.

— 1948: The origin of volutin on the chromosome, its transfer to the nucleolus, and suggestions concerning the significance of this phenomenon. Proc. nat. Acad. Sci. 34, 187—193.

— 1951: The relation of metaphosphate formation to cell division in yeast. Exper. Cell Res. 2, 275—278.

LISON, L., 1933: Sur les phenomenes de metachromasie. Bull. classe de sci. (Acad. roy. Belgique) 19, 1332—1341.

— 1935 a: Études sur la métachromasie. Colorants métachromatiques et substances chromotropes. Arch. Biol. (Fr.) 46, 599—668.

— 1935 b: La signification histochimique de la métachromasie. C. r. Soc. Biol. 118, 821—824.

— 1935 c: Sur la détermination du pH intracellulaire par les colorants vitaux indicateurs. L'erreur métachromatique. Protoplasma 24, 453—465.

— 1935 d: Sur la mécanisme et la signification de la coloration des lipides par le bleu de nil. Bull. Histol. appl. etc. 12, 279—289.

— 1936 a: Histochimie animale. Paris.

— 1936 b: Une réaction micro- et histochimique des esters sulfuriques complexes, la réaction métachromatique. Bull. Soc. Chim. biol. (Fr.) 18, 225—230.

— 1953: Histochimie et cytochimie animales. Paris. (2nd ed. of Histochimie animale.)

— and W. MUTSAARS, 1950: Metachromasy of nucleic acids. Quart. J. microsc. Sci. 91, 309—314.

LOEWI, G., 1953: Changes in the ground substance of aging cartilage. J. Path. a. Bacter. **65**, 381—388.

LOVE, R., and L. H. FROMMHAGEN, 1953: Histochemical studies on the clam, *Mactra solidissima*. Proc. Soc. exper. Biol. a. Med. (Am.) **83**, 838—844.

MACINTOSH, F. C., 1941: A colorimetric method for the standardization of heparin preparations. Biochem. J. **35**, 776—782.

MANCINI, R. E., 1950: Histochemical studies of techniques for mucopolysaccharides. J. nat. Cancer Inst. **10**, 1371.

MAST, S. O., and W. L. DOYLE, 1935 a: Structure, origin and function of cytoplasmic constituents in *Amoeba proteus*. I. Structure. Arch. Protistenk. **86**, 155—180.

— — 1935 b: Structure, origin and function of cytoplasmic constituents in *Amoeba proteus* with special reference to mitochondria and Golgi substance. II. Origin and function based on experimental evidence; effect of centrifuging on *Amoeba proteus*. Arch. Protistenk. **86**, 278—306.

MAXIMOW, A. A., and W. BLOOM, 1952: Textbook of Histology, 6th ed. Philadelphia.

McKAY, D. G., 1950: Metachromasia in the endometrium. Amer. J. Obstetr. **59**, 875—882.

MELLON, M. G., 1945: Colorimetry for Chemists. Columbus, Ohio.
— 1950: Analytical Absorption Spectroscopy. New York.

MERRILL, R. C., and R. W. SPENCER, 1948: Spectral changes of some dyes in soluble silicate solutions. J. amer. chem. Soc. **70**, 3683—3689.

— — and R. GETTY, 1948: The effect of sodium silicates on the absorption spectra of some dyes. J. amer. chem. Soc. **70**, 2460—2464.

MEYER, A., 1904: Orientierende Untersuchungen über Verbreitung, Morphologie und Chemie des Volutins. Bot. Z. **62**, 113—152.

MEYER, K., 1938: The chemistry and biology of mucopolysaccharides and glycoproteins. Cold Spring Harbor Symp. Quant. Biol. **6**, 91—102.
— 1947: The biological significance of hyaluronic acid and hyaluronidase. Physiol. Rev. (Am.) **27**, 335—359.
— 1951: Chemistry of connective tissue, polysaccharides. Trans. First Conf. on Connective Tissues, J. Macy Jr. Found., 88—100.
— 1953: Outline of problems to be solved in the study of connective tissues. Trans. Fourth Conf. on Connective Tissues, J. Macy Jr. Found., 185—197.
— and J. FELLIG, 1950: La constitution de l'acide hyaluronique. Experientia **6**, 186.
— A. LINKER, E. A. DAVIDSON, and B. WEISSMAN, 1953: The mucopolysaccharides of bovine cornea. J. biol. Chem. (Am.) **205**, 611—616.
— and M. M. RAPPORT, 1951: The mucopolysaccharides of the ground substance of connective tissue. Science **113**, 596—599.

MICHAELIS, L., 1903: Metachromasie. In EHRLICH Encykl. mikrosk. Technik **2**, 797—803.
— 1910: Metachromasie. In EHRLICH, Enzykl. mikrosk. Technik, 2nd ed., **2**, 77—82.
— 1926: Metachromasie. In KRAUSE, Enzykl. mikrosk. Technik, 3rd ed., **2**, 1376—1380.
— 1944: Theory of metachromatic staining. Biol. Bull. **87**, 155—156.
— 1947: The nature of the interaction of nucleic acids and nuclei with basic dyestuffs. Cold Spring Harbor Symp. Quant. Biol. **12**, 131—142.
— 1950: Reversible polymerization and molecular aggregation. J. Phys. a. Coll. Chem. **54**, 1—17.
— and S. GRANICK, 1945: Metachromasy of basic dyestuffs. J. amer. chem. Soc. **67**, 1212—1219.

MICHELS, N. A., 1938: The Mast Cells. In DOWNEY, Handbook of Hematology, Vol. I, 231—372.

MILLER, Z., B. J. WALDMAN, and F. C. McLEAN, 1952: The effect of dyes on the calcification of hypertrophic rachitic cartilage *in vitro*. J. exper. Med. (Am.) **95**, 497—508.

MINCHIN, E. A., 1909: The structure of *Trypanosoma lewisi* in relation to microscopical technique. Quart. J. microsc. Sci. **53**, 755—808.

MÖLLENDORFF, W. v., 1924: Untersuchungen zur Theorie der Färbung fixierter Präparate. III. WILH. und MILIE v. MÖLLENDORFF: Durchtränkungs- und Niederschlagsfärbung als Haupterscheinungen bei der histologischen Färbung. Erg. Anat. **25**, 1—66.

MONNÉ, L., and S. HÅRDE, 1951: On the cortical granules of the sea urchin egg. Ark. Zool. 1, 487—497.
— and D. B. SLAUTTERBACK, 1950: Differential staining of various polysaccharides in sea urchin eggs. Exper. Cell Res. 1, 477—491.
MONTAGNA, W., H. B. CHASE, and H. P. MELARAGNO, 1951: Histology and cyto-chemistry of human skin. I. Metachromasia in the mons pubis. J. nat. Cancer Inst. 12, 591—597.
— A. Z. EISEN, and A. S. GOLDMAN, 1954: The tinctorial behavior of human mast cells. Quart. J. microsc. Sci. 95, 1—4.
— and C. R. NOBACK, 1948: Localization of lipids and other chemical substances in the mast cells of man and laboratory animals. Anat. Rec. (Am.) 100, 535—545.
MOORE, J. E., III, and G. W. JAMES, III, 1953: A simple direct method for ab-solute basophil leucocyte count. Proc. Soc. exper. Biol. a. Med. (Am.) 82, 601—603.
MORGAN, W. T. J., 1949: Mucoids as components of the human erythrocyte sur-face. Exper. Cell Res., suppl. 1, 228—233.
MORI, T., 1953: Seaweed polysaccharides. Adv. Carboh. Chem. 8, 315—350.
MORRIONE, T. G., 1952: The formation of collagen fibers by the action of heparin on soluble collagen: an electron microscope study. J. exper. Med. (Am.) 96, 107—113.
MOSES, M. J., 1952: Quantitative optical techniques in the study of nuclear chemistry. Exper. Cell Res., suppl. 2, 75—94.
MOWRY, R. W., 1954: Histochemical demonstration and significance of dextran sulfate: a metachromatic, water-soluble acid polysaccharide. J. Histochem. a. Cytochem. 2, 470.
NAGEL, L., 1948: Volutin. Bot. Rev. 14, 178—184.
NEUBERG, C., 1949: Remarkable properties of nucleic acids and nucleotides. Arch. Biochem. 20, 185—210.
NEUMAN, W. F., E. S. BOYD, and I. FELDMAN, 1952: The ion binding properties of cartilage. Trans. Fourth Conf. on Metabolic Interrelations, J. Macy Jr. Found., 100—112.
NEUMANN, A., 1932: Chemie der Leukozyten. Handbuch der allg. Hämatol., H. HIRSCHFELD and A. HITTMAIR, Bd. I, 339—380.
NOBACK, C. R., 1954: Metachromasia in the nervous system. J. Neuropathol. a. exper. Neurol. 13, 161—167.
ODELL, L. D., 1952: Postpartum hemorrhage. Modern Med. Annual, 209—214. Minneapolis.
OHUYE, T., and T. H. MURAKAMI, 1953: On the metachromasy of a stain, Victoria blue 4 R. Memoirs Ehime Univ., Sec. II, 1, 339—343.
OLIVER, J., F. BLOOM, and C. MANGIERI, 1947: On the origin of heparin. An examination of the heparin content and the specific cytoplasmic particles of neoplastic mast cells. J. exper. Med. (Am.) 86, 107—116.
PAFF, G. H., and F. BLOOM, 1949: Vacuolation and the release of heparin in mast cells cultivated in vitro. Anat. Rec. (Am.) 104, 45—59.
PÁLOS, L. A., and CH. KOCSÁN, 1951: New data on the physico-chemical properties of heparin. Experientia 7, 97—98.
PAPPENHEIM, A., 1906: Allgemeine Leukocytologie der Entzündung. Theoretische Vorbemerkungen. Fol. haemat. (D.) 3, 564—569.
— 1910: Zusatz. Fol. haemat. (D.) 9, 642—643.
PEARSE, A. G. E., 1953: Histochemistry, Theoretical and Applied. Boston.
PENNEY, J. R., and B. M. BALFOUR, 1944: The effect of vitamin C on mucopoly-saccharide production in wound healing. J. Path. a. Bacter. 61, 171—178.
PERLMANN, G., 1938: On the preparation of crystallized egg albumin meta-phosphate. Biochem. J. 32, 931—932.
PETTERSSON, T., 1954: The effect of x-ray total-body irradiation on the mast cell count in the skin. Acta path. et microbiol. scand. (Dän.), suppl. 102, 62 pp.
PIKE, R. M., 1950: The production of hyaluronic acid and hyaluronidase by some strains of group A streptococci. Ann. N.Y. Acad. Sci. 52, 1070—1073.
PINCUS, P., 1950: A sulphated mucopolysaccharide in human dentine. Nature 166, 187.
POLLISTER, A. W., and L. ORNSTEIN, 1955: Cytophotometric analysis in the visible spectrum. In MELLORS, Analytical Cytology, Chap. 1, 1/3—1/71. New York.

Pollister, A. W., L. Ornstein, and H. Ris, 1947: Nucleoprotein determinations in cytological preparations. Cold Spring Harbor Symp. Quant. Biol. 12, 147—157.

Proescher, F., 1933: Pinacyanol as a histological stain. Proc. Soc. exper. Biol. a. Med. (Am.) 31, 79—81.

Rabinowitch, E., and L. F. Epstein, 1941: Polymerization of dyestuffs in solution. Thionine and methylene blue. J. amer. chem. Soc. 63, 69—78.

Ragan, C., 1951: Effect of ACTH and cortisone on connective tissues. Trans. First Conf. on Connective Tissues, J. Macy Jr. Found., 137—164.

— (ed.), 1953: General areas of agreement reached in this conference group. Trans. Fourth Conf. on Connective Tissue, J. Macy Jr. Found., 16—46.

Rawson, R. A., 1943: The binding of T-1824 and structurally related diazo dyes by the plasma proteins. Amer. J. Physiol. 138, 708—717.

Re, G., 1951: Ricerche istochimiche e biologiche sul muco degli involucri ovulari e degli ovidutti di anfibi. Arch. Biol. (Fr.), 62, 107—132 (Abstr. Biol. Abstracts 26, 3175, 1951).

Rienits, K. G., 1953: The electrophoresis of acid mucopolysaccharides on filter paper. Biochem. J. (Brit.) 53, 79—85.

Riley, J. F., 1948: Retardation of growth of a transplantable carcinoma in mice fed basic metachromatic dyes. Cancer Res. 8, 183—188.

— 1953: Histamine in tissue mast cells. Science 118, 332.

Roberts, H. S., and N. G. Anderson, 1951: Studies on isolated cell components. III. A cytological study of the effects of heparin on isolated nuclei. Exper. Cell Res. 2, 224—234.

Romanini, M. G., 1951: Contributions a l'étude histochimique des mucopolysaccharides. I⁰. Gelée de Wharton et substance métachromatique des vaisseaux. Acta anat. 13, 256—288.

Romeis, B., 1940: Hypophyse. In v. Möllendorff, Hdbch. mikrosk. Anat. d. Menschen 6, part III, 1—625.

Rubin, P. S., and J. E. Howard, 1950: Histochemical studies on the role of acid mucopolysaccharides in calcifiability and calcification. Trans. Second Conf. on Metabolic Interrelations, J. Macy Jr. Found., 155—166.

Runnström, J., 1949: The mechanism of fertilization in Metazoa. Adv. Enzymol. 9, 241—327.

— 1952: The cytoplasm, its structure and role in metabolism, growth and differentiation. In Barron, Modern Trends in Physiology and Biochemistry, 47—76.

Scheibe, G., 1938: Reversible Polymerisation als Ursache von neuartigen Absorptionsbändern von Farbstoffen. Kolloid-Z. 82, 1—14.

Schlechter, P., and M. Campani, 1948: Contributo allo studio della metachromasia. Arch. Sci. biol. (It.), 32, 166—182.

Schmidt, G., 1951: The biochemistry of inorganic pyrophosphates and metaphosphates. In McElroy and Glass, Phosphorus Metabolism 1, 443—476.

— L. Hecht, and S. J. Thannhauser, 1949: The effect of potassium ions on the absorption of orthophosphate and the formation of metaphosphate by baker's yeast. J. biol. Chem. (Am.) 178, 733—742.

Schubert, M., and A. Levine, 1953: A conductimetric study of the interaction of anionic mucopolysaccharides and cationic dyes. J. amer. chem. Soc. 75, 5842—5846.

Schulemann, W., 1915: Über Metachromasie bei Vitalfarbstoffen. Z. exper. Path. u. Ther. 17, 401—412.

Schwarz, R., and E. Hermann, 1922: Über die Metachromasie des Toluidinblaus. Kolloid-Z. 31, 91—94.

Sesachar, B. R., 1950: The nucleus and nucleic acids of Chilodonella uncinatus. J. exper. Zool. 114, 517—543.

— 1953: Metachromasy of the ciliate macronucleus. J. exper. Zool. 124, 117—127.

Sharp, L. W., 1943: Fundamentals of Cytology. New York.

Shear, M. J., and F. C. Turner, 1943: Chemical treatment of tumors. V. Isolation of the hemorrhage-producing fraction from Serratia marcescens (Bacillus prodigiosus) culture filtrate. J. nat. Cancer Inst. 4, 81—97.

Sheppard, S. E., and A. L. Geddes, 1944 a: Effect of solvents upon the absorption spectra of dyes. IV. Water as a solvent: a common pattern. J. amer. chem. Soc. 66, 1995—2002.

— — 1944 b: Effect of solvents upon the absorption spectra of dyes. V. Water as a solvent: quantitative examination of the dimerization hypothesis. J. amer. chem. Soc. 66, 2003—2009.

SIDMAN, R. L., and G. B. WISLOCKI, 1954: Histochemical observations on rods and cones in retinas of vertebrates. J. Histochem. a. Cytochem. 2, 413—433.

SINGER, M., 1952: Factors which control the staining of tissue sections with acid and basic dyes. Internat. Rev. Cytol. 1, 211—255.

— 1954: The staining of basophilic components. J. Histochem. a. Cytochem. 2. 322—333.

— and G. B. WISLOCKI, 1948: The affinity of syncytium, fibrin and fibrinoid of the human placenta for acid and basic dyes under controlled conditions of staining. Anat. Rec. (Am.) 102, 175—193.

SKUPIENSKI, F. X., 1929: Sur la coloration vitale de Didinium nigripes (Fr.) Acta Soc. Bot. Poloniae 3, 204—313 (abst. in Stain Techn. 6, 70—71, 1931).

SMITH, I. W., J. F. WILKINSON, and J. P. DUGUID, 1954: Volutin production in Aerobacter aerogenes due to nutrient imbalance. J. Bacter. (Am.) 68, 450—463.

SNELL, F. D., and C. T. SNELL, 1953: Colorimetric Methods of Analysis, 3rd ed. 3. New York.

SNELLMAN, O., R. JENSEN, and B. SYLVÉN, 1949: Notes on the fractionation and colorimetric assay of commercial heparin. Acta chem. scand. 3, 589—594.

— B. SYLVÉN, and C. JULÉN, 1951: Analysis of the native heparin-lipoprotein complex including the identification of a heparin complement (heparin co-factor) obtained from extracts of tissue mast cells. Biochim. et Biophysica Acta 7. 98—109.

SOBEL, A. E., 1955: Local factors in the mechanism of calcification. Ann. N.Y. Acad. Sci. 60, 713—732.

SOGNNAES, R. F., 1955: Microstructure and histochemical characteristics of the mineralized tissues. Ann. N.Y. Acad. Sci. 60, 545—574.

SPEIRS, R. S., 1955: Physiological approaches to an understanding of the function of eosinophils and basophils. Ann. N.Y. Acad. Sci. 59, 706—731.

SPEK, J., 1940: Metachromasie und Vitalfärbung mit pH-Indikatoren. Protoplasma 34, 533—584.

— 1942: Eine optische Methode zum Nachweis von Lipoiden in der lebenden Zelle. Protoplasma 37, 49—85.

— 1944: Optische Analysen von Vitalfärbungen. Jena. Z. Med. u. Naturw. 77, 48—67 (abst. Biol. Abst. 23, 2351, 1949).

STEDMAN, E., E. STEDMAN, and F. W. PETTIGREW, 1944: Retardation of growth of mouse carcinoma 2146 by histone and protamine. Biochem. J. (Brit.) 38, xxxi—xxxii.

STILLING, H., 1886: Fragmente zur Pathologie der Milz. Virchows Arch. 103, 15—38.

STOWELL, R. E., 1952: Use of histochemical and cytochemical technics in problems in pathology. Lab. Invest. 1, 210—230.

STRANDBERG, L., 1950: The preparation of chondroitin sulfuric acid. Acta physiol. scand. 21, 222—229.

SWELLENGREBEL, N.-H., 1908: La volutine chez les trypanosomes. C. r. Soc. Biol. 64, 38—40.

SYLVÉN, B., 1941: Über das Vorkommen von hochmolekularen Esterschwefelsäuren im Granulationsgewebe und bei der Epithelregeneration. Acta chir. scand., suppl. 66, 1—151.

— 1945: Ester sulphuric acids of high molecular weight and mast cells in mesenchymal tumors. Histochemical studies on tumorous growth. Acta radiol. (Schwd.) 59 (suppl.), 1—99.

— 1951: On the cytoplasmic constituents of normal tissue mast cells. Exper. Cell Res. 2, 252—255.

— 1954: Metachromatic dye-substrate interactions. Quart. J. microsc. Sci. 95, 327—358.

— and H. MALMGREN, 1952: On the alleged metachromasia of hyaluronic acid. Lab. Invest. 1, 413—431.

TAFT, E. B., 1951: The specificity of the methyl green-pyronin stain for nucleic acids. Exper. Cell Res. 2, 312—326.

TERAYAMA, H., 1949: On the nature of metachromasy. Jap. med. J. 2, 137—149.

— 1954: Application of the method of colloid titration to the study of bacteria. Arch. Biochem. a. Biophys. 50, 55—63.

THOMAS, L. J., Jr., 1951: A blood anti-coagulant from surf clams. Biol. Bull. (Am.) 101, 230.

— 1954: The localization of heparin-like blood anticoagulant substances in the tissues of Spisula solidissima. Biol. Bull. (Am.) 106, 129—138.

Tötterman, G., 1948: Are the basophilic leucocytes "heparinocytes"? Acta med. scand. (Schwd.) **130**, 176—182.

Tyler, F. A., 1954: Toluidine blue in the management of postoperative bleeding. Oral Surg., Oral Med., and Oral Path. **7**, 1066—1068.

Vasseur, E., 1948 a: Chemical studies on the jelly coat of the sea-urchin egg. Acta chem. scand. **2**, 900—913.

— 1948 b: The sulphuric acid content of the egg coat of the sea urchin, *Strongylocentrotus droebachiensis* Müll. Arkiv. Kemi, Mineral. och Geol. **25 B**, No. 6, 1—2.

Veen, G. van, and F. H. V. Meijer, 1948: Chemical analysis of a para-amyloid "tumour." Biochim. et Biophysica Acta **2**, 190—197.

Walton, K. W., and C. R. Ricketts, 1954: Investigation of the histochemical basis of metachromasia. Brit. J. exper. Path. **35**, 227—240.

Warren, G. H., 1950: The isolation of a mucopolysaccharide from *Aerobacter aerogenes*. Science **111**, 473—474.

Watson, J. D., and F. H. C. Crick, 1953: Molecular structure of nucleic acids. A structure for deoxyribose nucleic acid. Nature **171**, 737—738.

Weissman, N., W. H. Carnes, P. S. Rubin, and J. Fisher, 1952: Metachromasy of toluidine blue induced by nucleic acids. J. amer. chem. Soc. **74**, 1423—1426.

Whistler, R. 1..., and C. L. Smart, 1953: Polysaccharide Chemistry. New York.

Wiame, J. M.. 1946: Remarque sur la métachromasie des cellules de levure. C. r. Soc. Biol. **140**, 897—899.

— 1947 a: Étude d'une substance polyphosphorée, basophile et métachromatique chez les levures. Biochim. et Biophysica Acta **1**, 234—255.

— 1947 b: The metachromatic reaction of hexametaphosphate. J. amer. chem. Soc. **69**, 3146—3147.

— 1947 c: Yeast metaphosphate. Fed. Proc. **6**, 302.

— 1949: The occurrence and physiological behavior of two metaphosphate fractions in yeast. J. biol. Chem. (Am.) **178**, 919—929.

Wicklund, E., 1947: The action of clupein on the unfertilized sea-urchin eggs and its influence on the fertilization of these eggs. Arkiv Zool. **40 A**, no. 5.

Windle, W. F., R. Rhines, and J. Rankin, 1943: A Nissl method using buffered solutions of thionin. Stain Techn. **18**, 77—86.

Wislocki, G. B., 1950: Saliva-insoluble glycoproteins, stained by the periodic acid-Schiff procedure, in the placentas of pig, cat, mouse, rat, and man. J. nat. Cancer Inst. **10**, 1341.

— 1952: The anterior segment of the eye of the rhesus monkey investigated by histochemical means. Amer. J. Anat. **91**, 233—262.

— H. Bunting, and E. W. Dempsey, 1947 a: Further observations on the chemical cytology of megakaryocytes and other cells of hemopoietic tissues. Anat. Rec. (Am.) **98**, 527—537.

— — — 1947 b: Metachromasia in mammalian tissues and its relationship to mucopolysaccharides. Amer. J. Anat. **81**, 1—37.

— and E. W. Dempsey, 1948: The chemical histology of the human placenta and decidua with reference to mucopolysaccharides, glycogen, lipids and acid phosphatase. Amer. J. Anat. **83**, 1—41.

— and M. J. Singer, 1950: The basophilic and metachromatic staining of myelin sheaths and its possible association with a sulfatide. J. comp. Neurol. (Am.) **92**, 71—91.

— and R. F. Sognnaes, 1950: Histochemical reactions of normal teeth. Amer. J. Anat. **87**, 239—276.

Wolfrom, M. L., and W. H. McNeely, 1945: The relation between the structure of heparin and its anticoagulant activity. J. amer. chem. Soc. **67**, 748—753.

— D. I. Weisblat, J. V. Karabinos, W. H. McNeely, and J. McLean, 1943: Chemical studies on crystalline barium acid heparinate. J. amer. chem. Soc. **65**, 2077—2085.

Woodcock, H. M., 1906: The hemoflagellates: A review of present knowledge relating to the trypanosomes and allied forms. Quart. J. microsc. Sci. **50**, 151—332.

Worley, L. M., and V. S. Lequire, 1955: Lipemia clearing in peptone and anaphylactic shock. Proc. Soc. exper. Biol. a. Med. (Am.) **89**, 181—183.

Yamamoto, T., 1952: On the cortical changes of the unfertilized eggs of *Tylorrhincus heterochaetus* at the time of fertilization and artificial activation, with special reference to the properties of the cortical granules. Zikkenseibutsugakuho **2**, 193 (in Japanese, cited from Kusa, 1954).

Yasuzumi, G., T. Mori, H. Matsukura, and R. Minamino, 1950: Histochemical studies on chromosomes. Cytologia 15, 173—182.
Ziff, M., and E. Chargaff, 1940: Studies on the chemistry of blood coagulation. XI. The mode of action of heparin. J. biol. Chem. (Am.) 136, 689—695.

Addendum

Braden, A. W. H., 1952: Properties of the membranes of rat and rabbit eggs. Austral. J. Sci. Res. (B) 5, 460—471. (Histochemical study.)
Dalcq, A. M., 1955: Processes of synthesis during early development of rodents' eggs and embryos. Studies on Fertility 7, 113—122. (Mucopolysaccharides as precursors of chromosomal materials.)
Pasteels, J., 1955: Évolution de la métachromasie au bleu de toluidine suivie sur le vivant, au cours des premières phases du développement de l'Oursin et de la Pholade. Bull. Acad. roy. belg. (cl. de sci.) 41, 761—768. (Perinuclear metachromasy related to mitochondria.)
Schubert, M., and A. Levine, 1955: A qualitative theory of metachromasy in solution. J. Am. Chem. Soc. 77, 4197—4201. (Suggests possibility of a quantitative theory.)
— and D. Hamerman, 1956: Metachromasia; chemical theory and histochemical use. J. Histochem. and Cytochem. 4, 159—189. (Review.)
Spek, J., and G. Gilissen, 1943: Die Zellmembran der Amoeben — eine chromotrope Substanz. Protoplasma 37, 258—272. (Vital staining, "Fällungsmetachromasie", mucoids in cell surface.)
Woohsman, H., 1956 a: Der Stand des Metachromasie-Problems. Protoplasma 45, 618—629. (Short review: relation of metachromasy to mechanism of "Einschlußverbindungen".)
— 1956 b: Die Fällungsmetachromasie der chromotropen Substanzen der Ciliaten and Flagellaten. Protoplasma 47, 37—66.

SPRINGER-VERLAG IN WIEN I

Protoplasmatologia. Handbuch der Protoplasmaforschung

Herausgegeben von

L. V. Heilbrunn, Philadelphia, und **F. Weber,** Graz

Das Handbuch erscheint in selbständigen Einzelveröffentlichungen, die zu 14 Bänden vereinigt werden. Jeder selbständig erscheinende Handbuchteil ist einzeln käuflich. Bei Verpflichtung zur Abnahme des gesamten Handbuches, bei Vorbestellung der einzelnen Teile sowie für Abonnenten der Zeitschrift „Protoplasma" ermäßigt sich der Preis um 20%

Bisher sind erschienen:

Die makromolekulare Chemie und ihre Bedeutung für die Protoplasmaforschung. Von Prof. Dr. phil., Dr.-Ing. e. h., Dr. rer. nat. h. c., Dr. (C) h. c. **Hermann Staudinger** und Dr. phil., Mag. rer. nat. **Magda Staudinger,** beide Staatliches Forschungsinstitut für makromolekulare Chemie der Universität Freiburg i. Br. **Band I. Grundlagen.** 1. Die makromolekulare Chemie und ihre Bedeutung für die Protoplasmaforschung. Mit 27 Textabbildungen. IV, 73 Seiten. Gr.-8°. 1954.
S 117.—, DM 19.50, sfr. 20.—, $ 4.65

Die submikroskopische Struktur des Cytoplasmas. Von Prof. Dr. **A. Frey-Wyssling,** Institut für Allgemeine Botanik der Eidg. Technischen Hochschule Zürich. **Band II. Cytoplasma.** A. Morphologie. 2. Die submikroskopische Struktur des Cytoplasmas. Mit 90 Textabbildungen. IV, 244 Seiten. Gr.-8°. 1955.
S 255.—, DM 42.50, sfr. 43.50, $ 10.10

The pH of Plant Cells. By Prof. Dr. **James Small,** The Queen's University Belfast, Department of Botany. With 3 figures. 116 pages. — **The pH of Animal Cells.** By Professor Dr. **Floyd J. Wierclnski,** Hahnemann Medical College, Department of Physiology, Philadelphia, Pa. With 7 figures. 56 pages. **Band II. Cytoplasma.** B. Chemie. 2. Spezielle Cytochemie und Histochemie. c. The pH of Plant Cells. The pH of Animal Cells. Gr.-8°. 1955.
S 270.—, DM 45.—, sfr. 46.—, $ 10.70

The Enzymology of the Cell Surface. By **Aser Rothstein,** Rochester, New York. With 21 figures. 86 pages. — **Tension at the Cell Surface.** By **E. Newton Harvey,** Princeton, New Jersey. With 13 figures. 30 pages. **Band II. Cytoplasma.** E. Cytoplasma-Oberfläche. 4. The Enzymology of the Cell Surface. 5. Tension at the Cell Surface. Gr.-8°. 1954.
S 168.—, DM 28.—, sfr. 28.80, $ 6.70

Chemistry and Physiology of Mitochondria and Microsomes. By **Olov Lindberg,** Ph. D., and **Lars Ernster,** Ph. D., beide Wenner-Gren's Institute, Stockholm. **Band III. Cytoplasma-Organellen.** A. Chondriosomen, Mikrosomen, Sphaerosomen. 4. Chemistry and Physiology of Mitochondria and Microsomes. With 32 figures. IV, 136 pages. Gr.-8°. 1954.
S 204.—, DM 34.—, sfr. 34.80, $ 8.10

Endomitose und endomitotische Polyploidisierung. Von Prof. Dr. **Lothar Geitler,** Botanisches Institut der Universität Wien. **Band VI. Kern- und Zellteilung.** C. Endomitose und endomitotische Polyploidisierung. Mit 44 Textabbildungen. IV, 89 Seiten. Gr.-8°. 1953.
S 140.—, DM 23.50, sfr. 24.10, $ 5.60

Active Transport through Animal Cell Membranes. By Dr. **Paul G. LeFevre,** Medical Branch, Division of Biology and Medicine, United States Atomic Energy Commission, Washington, D. C. **Band VIII. Physiologie des Protoplasmas.** 7. Aktiver Stofftransport. a. Active Transport through Animal Cell Membranes. With 31 figures. IV, 123 pages. Gr.-8°. 1955.
S 228.—, DM 38.—, sfr. 38.70, $ 9.—

Red Cell Structure and Its Breakdown. By Prof. Dr. **Eric Ponder,** The Nassau Hospital, Mineola, N.Y. **Band X. Pathologie des Protoplasmas.** 2. Red Cell Structure and Its Breakdown. With 58 figures. IV, 123 pages. Gr.-8°. 1955.
S 240.—, DM 40.—, sfr. 40.90, $ 9.50

Protoplasmatische Pflanzenanatomie. Von Dr. **Lotte Reuter,** Privatdozent an der Universität Wien. **Band XI. Vergleichende Protoplasmatik.** 2. Protoplasmatische Pflanzenanatomie. Mit 64 Textabbildungen. IV, 131 Seiten. Gr.-8°. 1955.
S 204.—, DM 34.—, sfr. 34.80, $ 8.10

Zu beziehen durch jede Buchhandlung

PROTOPLASMA

Unter besonderer Mitwirkung von

R. Chambers   N. Kamiya   S. Strugger
New York   Osaka   Münster

herausgegeben von

J. Spek   F. Weber   K. Höfler
Rostock   Graz   Wien

Springer-Verlag in Wien

Die Zeitschrift erscheint zwanglos in einzeln berechneten Heften wechselnden Umfanges, die zu Bänden vereinigt werden. Über Bezugsbedingungen, Preise, Inhalt der erschienenen Hefte usw. erteilt der Verlag bereitwilligst Auskunft.

Von der 1926 gegründeten Zeitschrift sind trotz einer kriegsbedingten Unterbrechung von sechs Jahren bis 1956 46 Bände erschienen. Der Kreis ihrer Mitarbeiter und die veröffentlichten Arbeiten der letzten Jahre beweisen, daß sie heute wieder jene führende Stellung einnimmt, die sie vor dem Kriege hatte, und ihre Aufgabe voll erfüllt: eine wahrhaft internationale Sammelstelle aller Forschungsarbeiten zu sein, die sich mit dem Lebenssubstrat befassen, ob dieses nun pflanzlicher, tierischer oder menschlicher Herkunft ist.

Zu beziehen durch jede Buchhandlung